Instructor's Manual

for

Elements of Modern Algebra

Sixth Edition

Jimmie Gilbert
University of South Carolina Upstate

Linda Gilbert
University of South Carolina Upstate

Australia • Canada • Mexico • Singapore • Spain • United Kingdom • United States

Printed in the United States of America
1 2 3 4 5 6 7 08 07 06 05 04

Printer: Malloy Incorporated

ISBN: 0-534-40263-1

For more information about our products, contact us at:
Thomson Learning Academic Resource Center
1-800-423-0563

For permission to use material from this text or product, submit a request online at **http://www.thomsonrights.com.**
Any additional questions about permissions can be submitted by email to **thomsonrights@thomson.com.**

Thomson Brooks/Cole
10 Davis Drive
Belmont, CA 94002-3098
USA

Asia
Thomson Learning
5 Shenton Way #01-01
UIC Building
Singapore 068808

Australia/New Zealand
Thomson Learning
102 Dodds Street
Southbank, Victoria 3006
Australia

Canada
Nelson
1120 Birchmount Road
Toronto, Ontario M1K 5G4
Canada

Europe/Middle East/South Africa
Thomson Learning
High Holborn House
50/51 Bedford Row
London WC1R 4LR
United Kingdom

Latin America
Thomson Learning
Seneca, 53
Colonia Polanco
11560 Mexico D.F.
Mexico

Spain/Portugal
Paraninfo
Calle/Magallanes, 25
28015 Madrid, Spain

Contents

Preface

This manual provides answers for the computational exercises and a few of the exercises requiring proofs in **Elements of Modern Algebra,** *Sixth Edition,* by Jimmie Gilbert and Linda Gilbert. These exercises are listed in the table of contents. In constructing proof of exercises, we have freely utilized prior results, including those results stated in preceding problems.

Our sincere thanks go to Stacy Green for her careful management of the production of this manual and to Eric T. Howe for his excellent work on the accuracy checking of all the answers.

Jimmie Gilbert
Linda Gilbert

Answers to Selected Exercises

Exercises 1.1

1. a. $A = \{x \mid x \text{ is a nonnegative even integer less than } 12\}$ b. $\{x \mid x^2 = 1\}$
   c. $A = \{x \mid x \text{ is a negative integer}\}$ d. $\{x \mid x = n^2 \text{ for } n \in \mathbf{Z}^+\}$

2. a. False b. True c. False d. False e. False f. True

3. a. False b. True c. True d. False e. True f. True
   g. False h. False i. False j. True

4. a. True b. False c. False d. True e. False f. False
   g. False h. False

5. a. $\{0,1,2,3,4,5,6,8,10\}$ b. $\{2,3,5\}$ c. $\{0,2,4,6,7,8,9,10\}$ d. $\{2\}$
   e. $\varnothing$ f. A g. $\{0,2,3,4,5\}$ h. $\{6,8,10\}$ i. $\{1,3,5\}$
   j. $\{6,8,10\}$ k. $\{1,2,3,5\}$ l. C m. $\{3,5\}$ n. $\{1\}$

6. a. A b. A c. $\varnothing$ d. U e. A f. $\varnothing$ g. A h. U i. U
   j. A k. U l. $\varnothing$ m. A n. $\varnothing$

7. a. $\{\varnothing, A\}$ b. $\{\varnothing, \{0\}, \{1\}, A\}$
   c. $\{\varnothing, \{a\}, \{b\}, \{c\}, \{a,b\}, \{a,c\}, \{b,c\}, A\}$
   d. $\{\varnothing, \{1\}, \{2\}, \{3\}, \{4\}, \{1,2\}, \{1,3\}, \{1,4\}, \{2,3\}, \{2,4\}, \{3,4\}, \{1,2,3\}, \{1,2,4\}, \{1,3,4\}, \{2,3,4\}, A\}$
   e. $\{\varnothing, \{1\}, \{\{1\}\}, A\}$ f. $\{\varnothing, A\}$ g. $\{\varnothing, A\}$ h. $\{\varnothing, \{\varnothing\}, \{\{\varnothing\}\}, A\}$

8. a. Two possible partitions are:
   $X_1 = \{x \mid x \text{ is a negative integer}\}$ and $X_2 = \{x \mid x \text{ is a nonnegative integer}\}$, or
   $X_1 = \{x \mid x \text{ is a negative integer}\}, X_2 = \{x \mid x \text{ is a positive integer}\}, X_3 = \{0\}$.
   b. One possible partition is $X_1 = \{a,b\}$ and $X_2 = \{c,d\}$. Another possible partition is $X_1 = \{a\}, X_2 = \{b,c\}, X_3 = \{d\}$.

c. One partition is $X_1 = \{1,5,9\}$ and $X_2 = \{11,15\}$. Another partition is $X_1 = \{1,15\}, X_2 = \{11\}$ and $X_3 = \{5,9\}$.

d. One possible partition is $X_1 = \{x \mid x = a + bi$, where a is a positive real number, b is a real number$\}$ and $X_2 = \{x \mid x = a + bi$, where a is a nonpositive real number, b is a real number$\}$. Another possible partition is $X_1 = \{x \mid x = a$, where a is a real number$\}$, $X_2 = \{x \mid x = bi$, where b is a nonzero real number$\}$ and $X_3 = \{x \mid x = a + bi$, where a and b are both nonzero real numbers$\}$.

9. **a**. $X_1 = \{1\}, X_2 = \{2\}, X_3 = \{3\}$; $X_1 = \{1\}, X_2 = \{2,3\}$; $X_1 = \{2\}, X_2 = \{1,3\}$; $X_1 = \{3\}, X_2 = \{1,2\}$

b. $X_1 = \{1\}, X_2 = \{2\}, X_3 = \{3\}, X_4 = \{4\}$;
$X_1 = \{1\}, X_2 = \{2\}, X_3 = \{3,4\}$; $X_1 = \{1\}, X_2 = \{3\}, X_3 = \{2,4\}$;
$X_1 = \{1\}, X_2 = \{4\}, X_3 = \{2,3\}$; $X_1 = \{2\}, X_2 = \{3\}, X_3 = \{1,4\}$;
$X_1 = \{2\}, X_2 = \{4\}, X_3 = \{1,3\}$; $X_1 = \{3\}, X_2 = \{4\}, X_3 = \{1,2\}$;
$X_1 = \{1,2\}, X_2 = \{3,4\}$; $X_1 = \{1,3\}, X_2 = \{2,4\}$;
$X_1 = \{1,4\}, X_2 = \{2,3\}$; $X_1 = \{1\}, X_2 = \{2,3,4\}$;
$X_1 = \{2\}, X_2 = \{1,3,4\}$; $X_1 = \{3\}, X_2 = \{1,2,4\}$;
$X_1 = \{4\}, X_2 = \{1,2,3\}$.

10. 2^n **11**. $A = \{a,b\}, B = \{a,c\}$

12. **a**. $A \subseteq B$ **b**. $B' \subseteq A$ or $A \cup B = U$ **c**. $B \subseteq A$
d. $A \cap B = \varnothing$ or $A \subseteq B'$ **e**. $A = B = U$ **f**. $A' \subseteq B$ or $A \cup B = U$
g. $A = U$ **h**. $A = U$

35. $(A \cap B') \cup (A' \cap B) = (A \cup B) \cap (A' \cup B')$

36. **a**.

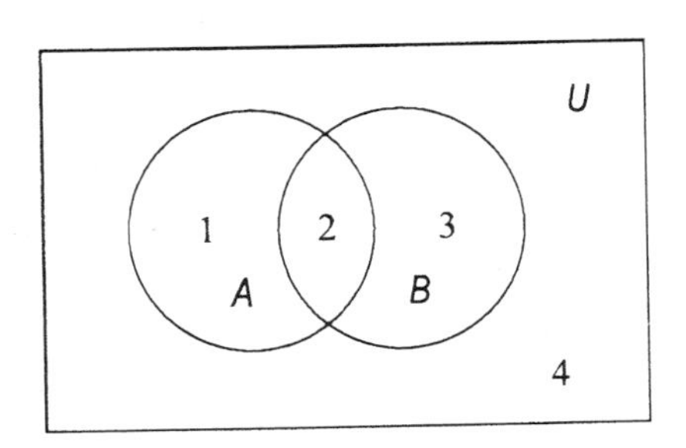

$A \cup B$: Regions 1,2,3

$A \cap B$: Region 2

$(A \cup B) - (A \cap B)$: Regions 1,3

$A + B$: Regions 1,3

$A - B$: Region 1

$B - A$: Region 3

$(A - B) \cup (B - A)$: Regions 1,3

Each of $A + B$ and $(A - B) \cup (B - A)$ consists of Regions 1,3.

b.

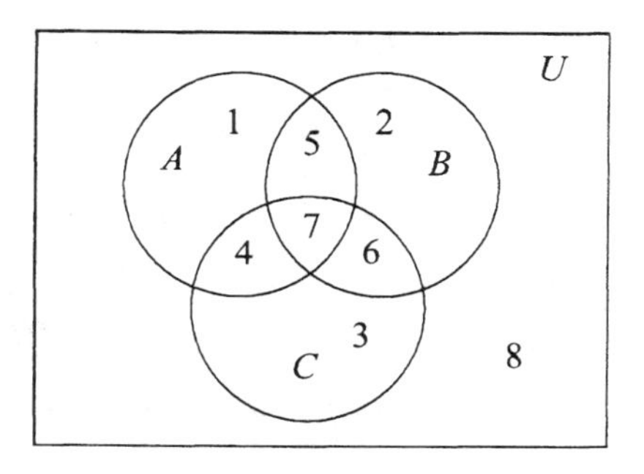

A : Regions 1,4,5,7	$A+B$: Regions 1,2,4,6
$B+C$: Regions 2,3,4,5	C : Regions 3,4,6,7
$A+(B+C)$: Regions 1,2,3,7	$(A+B)+C$: Regions 1,2,3,7

Each of $A+(B+C)$ and $(A+B)+C$ consists of Regions 1,2,3,7.

c.

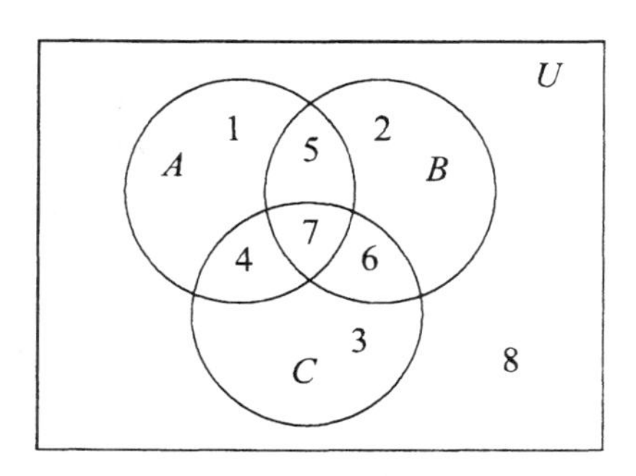

A : Regions 1,4,5,7	$A \cap B$: Regions 5,7
$B+C$: Regions 2,3,4,5	$A \cap C$: Regions 4,7
$A \cap (B+C)$: Regions 4,5	$(A \cap B)+(A \cap C)$: Regions 4,5

Each of $A \cap (B+C)$ and $(A \cap B)+(A \cap C)$ consists of Regions 4,5.

Exercises 1.2

1. **a**. $\{(a,0),(a,1),(b,0),(b,1)\}$ **b**. $\{(0,a),(0,b),(1,a),(1,b)\}$

c. $\{(2,2),(4,2),(6,2),(8,2)\}$

d. $\{(-1,1),(-1,5),(-1,9),(1,1),(1,5),(1,9)\}$

e. $\{(1,1),(1,2),(1,3),(2,1),(2,2),(2,3),(3,1),(3,2),(3,3)\}$

2. **a**. Domain $= \mathbf{E}$, Codomain $= \mathbf{Z}$, Range $= \mathbf{Z}$

b. Domain $= \mathbf{E}$, Codomain $= \mathbf{Z}$, Range $= \mathbf{E}$

c. Domain $= \mathbf{E}$, Codomain $= \mathbf{Z}$, Range $= \{y \mid y$ is a nonegative even integer$\} = (\mathbf{Z}^+ \cap \mathbf{E}) \cup \{0\}$

d. Domain $= \mathbf{E}$, Codomain $= \mathbf{Z}$, Range $= \mathbf{Z} - \mathbf{E}$

3. a. $f(S) = \{1, 3, 5, \ldots\} = \mathbf{Z}^+ - \mathbf{E}, f^{-1}(T) = \{-4, -3, -1, 1, 3, 4\}$

b. $f(S) = \{1, 5, 9\}, f^{-1}(T) = \mathbf{Z}$ c. $f(S) = \{0, 1, 4\}, f^{-1}(T) = \varnothing$

d. $f(S) = \{0, 2, 14\}, f^{-1}(T) = \mathbf{Z}^+ \cup \{0, -1, -2\}$

4. a. The mapping f is not onto, since there is no $x \in \mathbf{Z}$ such that $f(x) = 1$. It is one-to-one.

b. The mapping f is not onto since there is no $x \in \mathbf{Z}$ such that $f(x) = 1$. It is one-to-one.

c. The mapping f is onto and one-to-one.

d. The mapping f is one-to-one. It is not onto since there is no $x \in \mathbf{Z}$ such that $f(x) = 2$.

e. The mapping f is not onto, since there is no $x \in \mathbf{Z}$ such that $f(x) = -1$. It is not one-to-one, since $f(1) = f(-1)$ and $1 \neq -1$.

f. We have $f(3) = f(2) = 0$, so f is not one-to-one. Since $f(x)$ is always even, there is no $x \in \mathbf{Z}$ such that $f(x) = 1$, and f is not onto.

g. The mapping f is not onto, since there is no $x \in \mathbf{Z}$ such that $f(x) = 3$. It is one-to-one.

h. The mapping f is not onto since there is no $x \in \mathbf{Z}$ such that $f(x) = 1$. Neither is f one-to-one since $f(0) = f(1)$ and $0 \neq 1$.

i. The mapping f is onto. It is not one-to-one, since $f(9) = f(4)$ and $9 \neq 4$.

j. The mapping f is not onto since there is no $x \in \mathbf{Z}$ such that $f(x) = 4$. It is one-to-one.

5. a. The mapping is onto and one-to-one.

b. The mapping is onto and one-to-one.

c. The mapping is onto and one-to-one.

d. The mapping is onto and one-to-one.

e. The mapping is not onto since there is no $x \in \mathbf{R}$ such that $f(x) = -1$. It is not one-to-one since $f(1) = f(-1)$ and $1 \neq -1$.

f. The mapping is not onto since there is no $x \in \mathbf{R}$ such that $f(x) = 1$. It is not one-to-one since $f(0) = f(1) = 0$ and $0 \neq 1$.

6. a. The mapping f is onto and one-to-one.

b. The mapping f is one-to-one. Since there is no $x \in \mathbf{E}$ such that $f(x) = 2$, the mapping is not onto.

7. a. The mapping f is onto. The mapping f is not one-to-one since $f(1) = f(-1)$ and $1 \neq -1$.

   b. The mapping f is not onto since there is no $x \in \mathbf{Z}^+$ such that $f(x) = -1$. The mapping f is one-to-one.

   c. The mapping f is onto and one-to-one.

   d. The mapping f is onto. The mapping f is not one-to-one since $f(1) = f(-1)$ and $1 \neq -1$.

8. a. Let $f : \mathbf{E} \to \mathbf{E}$ where $f(x) = x$. b. Let $f : \mathbf{E} \to \mathbf{E}$ where $f(x) = 2x$.

   c. Let $f : \mathbf{E} \to \mathbf{E}$ where $f(x) = \begin{cases} \frac{x}{2} & \text{if } x \text{ is a multiple of } 4 \\ x & \text{if } x \text{ is not a multiple of } 4. \end{cases}$

   d. Let $f : \mathbf{E} \to \mathbf{E}$ where $f(x) = x^2$.

9. a. For arbitrary a in $\mathbf{Z}, 2a - 1$ is odd, and therefore

$$f(2a-1) = \frac{(2a-1)+1}{2} = a.$$

   Thus f is onto. But f is not one-to-one, since $f(2) = 5$ and also $f(9) = 5$.

   b. For arbitrary a in $\mathbf{Z}$, $2a$ is even and $f(2a) = \frac{2a}{2} = a$. Thus f is onto. But f is not one-to-one since $f(4) = 2$ and $f(7) = 2$.

   c. The mapping f is not onto, because there is no x in $\mathbf{Z}$ such that $f(x) = 4$. Since $f(2) = 6$ and $f(3) = 6$, then f is not one-to-one.

   d. The mapping f is not onto since there is no x in $\mathbf{Z}$ such that $f(x) = 1$. Suppose that $f(a_1) = f(a_2)$. It can be seen from the definition of f that the image of an even integer is always an odd integer, and also that the image of an odd integer is always an even integer. Therefore $f(a_1) = f(a_2)$ requires that either both a_1 and a_2 are even, or both a_1 and a_2 are odd. If both a_1 and a_2 are even,

$$f(a_1) = f(a_2) \Rightarrow 2a_1 - 1 = 2a_2 - 1 \Rightarrow 2a_1 = 2a_2 \Rightarrow a_1 = a_2.$$

   If both a_1 and a_2 are odd,

$$f(a_1) = f(a_2) \Rightarrow 2a_1 = 2a_2 \Rightarrow a_1 = a_2.$$

   Hence $f(a_1) = f(a_2)$ always implies $a_1 = a_2$ and f is one-to-one.

10. a. The mapping f is not onto, because there is no $x \in \mathbf{R} - \{0\}$ such that

$f(x) = 1$. If $a_1, a_2 \in \mathbf{R} - \{0\}$,

$$\begin{aligned} f(a_1) = f(a_2) \Rightarrow & \quad \frac{a_1 - 1}{a_1} = \frac{a_2 - 1}{a_2} \\ \Rightarrow & \quad a_2(a_1 - 1) = a_1(a_2 - 1) \\ \Rightarrow & \quad a_2 a_1 - a_2 = a_1 a_2 - a_1 \\ \Rightarrow & \quad -a_2 = -a_1 \\ \Rightarrow & \quad a_2 = a_1. \end{aligned}$$

Thus f is one-to-one.

b. The mapping f is not onto, because there is no $x \in \mathbf{R} - \{0\}$ such that $f(x) = 2$. If $a_1, a_2 \in \mathbf{R} - \{0\}$,

$$\begin{aligned} f(a_1) = f(a_2) \Rightarrow & \quad \frac{2a_1 - 1}{a_1} = \frac{2a_2 - 1}{a_2} \\ \Rightarrow & \quad 2 - \frac{1}{a_1} = 2 - \frac{1}{a_2} \\ \Rightarrow & \quad -\frac{1}{a_1} = -\frac{1}{a_2} \\ \Rightarrow & \quad a_1 = a_2 \end{aligned}$$

Thus f is one-to-one.

c. The mapping f is not onto, since there is no $x \in \mathbf{R} - \{0\}$ such that $f(x) = 0$. It is not one-to-one, since $f(2) = \frac{2}{5}$ and $f\left(\frac{1}{2}\right) = \frac{2}{5}$.

d. The mapping f is not onto since there is no $x \in \mathbf{R} - \{0\}$ such that $f(x) = 1$. Since $f(1) = f(3) = \frac{1}{2}$, then f is not one-to-one.

11. a. The mapping f is onto, since for every $(y, x) \in B = \mathbf{Z} \times \mathbf{Z}$ there exists an $(x, y) \in A = \mathbf{Z} \times \mathbf{Z}$ such that $f(x, y) = (y, x)$. To show that f is one-to-one, we assume $(a, b) \in A = \mathbf{Z} \times \mathbf{Z}$ and $(c, d) \in A = \mathbf{Z} \times \mathbf{Z}$ and

$$f(a, b) = f(c, d)$$

or

$$(b, a) = (d, c).$$

This means $b = d$ and $a = c$ and

$$(a, b) = (c, d).$$

b. For any $x \in \mathbf{Z}$, $(x, 0) \in A$ and $f(x, 0) = x$. Thus f is onto. Since $f(2, 3) = f(4, 1) = 5$, f is not one-to-one.

c. Since for every $x \in B = \mathbf{Z}$ there exists an $(x,y) \in A = \mathbf{Z} \times \mathbf{Z}$ such that $f(x,y) = x$, the mapping f is onto. However, f is not one-to-one, since $f(1,0) = f(1,1)$ and $(1,0) \neq (1,1)$.

d. The mapping f is one-to-one since $f(a_1) = f(a_2) \Rightarrow (a_1, 1) = (a_2, 1) \Rightarrow a_1 = a_2$. Since there is no $x \in \mathbf{Z}$ such that $f(x) = (0,0)$, then f is not onto.

12. a. The mapping f is not onto, since there is no $a \in A$ such that $f(a) = 9 \in B$. It is not one-to-one, since $f(-2) = f(2)$ and $-2 \neq 2$.

b. $f^{-1}(f(S)) = f^{-1}(\{1,4\}) = \{-2,1,2\} \neq S$

c. With $T = \{4,9\}$, $f^{-1}(T) = \{-2,2\}$, and $f\left(f^{-1}(T)\right) = f(\{-2,2\}) = \{4\} \neq T$.

13. a. $g(S) = \{2,4\}, g^{-1}(g(S)) = \{2,3,4,7\}$

b. $g^{-1}(T) = \{9,6,11\}, g\left(g^{-1}(T)\right) = T$

14. a. $f(S) = \{-1,2,3\}, f^{-1}(f(S)) = S$ b. $f^{-1}(T) = \{0\}, f\left(f^{-1}(T)\right) = \{-1\}$

15. a. $(f \circ g)(x) = \begin{cases} 2x & \text{if } x \text{ is even} \\ 2(2x-1) & \text{if } x \text{ is odd} \end{cases}$ b. $(f \circ g)(x) = 2x^3$

c. $(f \circ g)(x) = \begin{cases} \dfrac{x+|x|}{2} & \text{if } x \text{ is even} \\ |x| - x & \text{if } x \text{ is odd} \end{cases}$ d. $(f \circ g)(x) = x$

e. $(f \circ g)(x) = (x - |x|)^2$

16. a $(g \circ f)(x) = 2x$ b. $(g \circ f)(x) = 8x^3$ c. $(g \circ f)(x) = \dfrac{x+|x|}{2}$

d. $(g \circ f)(x) = \begin{cases} \frac{x}{2} - 1, & \text{if } x = 4k, \text{ for } k \text{ an integer} \\ x, & \text{otherwise} \end{cases}$ e. $(g \circ f)(x) = 0$

17. n^m **18.** $n!$ **19.** $n(n-1)(n-2)\cdots(n-m+1) = \dfrac{n!}{(n-m)!}$

25. Let $f : A \to B$, where A and B are nonempty.

Assume first that $f\left(f^{-1}(T)\right) = T$ for every subset T of B. For an arbitrary element b of B, let $T = \{b\}$. The equality $f\left(f^{-1}(\{b\})\right) = \{b\}$ implies that $f^{-1}(\{b\})$ is not empty. For any $x \in f^{-1}(\{b\})$, we have $f(x) = b$. Thus f is onto.

Assume now that f is onto. For an arbitrary $y \in f\left(f^{-1}(T)\right)$, we have

$$\begin{aligned} y \in f\left(f^{-1}(T)\right) &\Rightarrow y = f(x) \text{ for some } x \in f^{-1}(T) \\ &\Rightarrow y = f(x) \text{ for some } f(x) \in T \\ &\Rightarrow y \in T. \end{aligned}$$

Thus $f\left(f^{-1}(T)\right) \subseteq T$. For an arbitrary $t \in T$, there exists $x \in A$ such that $f(x) = t$, since f is onto. Now

$$\begin{aligned} f(x) = t \in T &\Rightarrow x \in f^{-1}(T) \\ &\Rightarrow f(x) \in f\left(f^{-1}(T)\right) \\ &\Rightarrow t \in f\left(f^{-1}(T)\right). \end{aligned}$$

Thus $T \subseteq f\left(f^{-1}(T)\right)$, and we have proved that $f\left(f^{-1}(T)\right) = T$ for an arbitrary subset T of B.

Exercises 1.3

1. a. The mapping $f \circ g$ is not onto since there is no $x \in \mathbf{Z}$ such that $(f \circ g)(x) = 1$. The mapping $f \circ g$ is one-to-one.

   b. The mapping $f \circ g$ is not onto since there is no $x \in \mathbf{Z}$ such that $(f \circ g)(x) = 1$. The mapping $f \circ g$ is one-to-one.

   c. The mapping $f \circ g$ is not onto since there is no $x \in \mathbf{Z}$ such that $(f \circ g)(x) = 1$. It is not one-to-one, since $(f \circ g)(-2) = (f \circ g)(0)$ and $-2 \neq 0$.

   d. The mapping $f \circ g$ is both onto and one-to-one.

   e. The mapping $f \circ g$ is not onto, since there is no $x \in \mathbf{Z}$ such that $(f \circ g)(x) = -1$. It is not one-to-one, since $(f \circ g)(1) = (f \circ g)(2)$ and $1 \neq 2$.

2. a. The mapping $g \circ f$ is not onto, since there is no $x \in \mathbf{Z}$ such that $(g \circ f)(x) = 1$. The mapping $g \circ f$ is one-to-one.

   b. The mapping $g \circ f$ is not onto, since there is no $x \in \mathbf{Z}$ such that $(g \circ f)(x) = 1$. The mapping $g \circ f$ is one-to-one.

   c. The mapping $g \circ f$ is not onto, since there is no $x \in \mathbf{Z}$ such that $(g \circ f)(x) = -1$. It is not one-to-one, since $(g \circ f)(-1) = (g \circ f)(-2)$ and $-1 \neq -2$.

   d. The mapping $g \circ f$ is not onto, since there is no $x \in \mathbf{Z}$ such that $(g \circ f)(x) = 0$. The mapping $g \circ f$ is not one-to-one, since $(g \circ f)(1) = (g \circ f)(4)$ and $1 \neq 4$.

   e. The mapping $g \circ f$ is not onto, since there is no $x \in \mathbf{Z}$ such that $(g \circ f)(x) = 1$. It is not one-to-one, since $(g \circ f)(0) = (g \circ f)(1)$ and $0 \neq 1$.

3. Let $A = \{0, 1\}, B = \{-2, 1, 2\}, C = \{1, 4\}$. Let $g : A \to B$ be defined by $g(x) = x+1$ and $f : B \to C$ be defined by $f(x) = x^2$. Then g is not onto, since $-2 \notin g(A)$. The mapping f is onto. Also $f \circ g$ is onto, since $(f \circ g)(0) = f(1) = 1$ and $(f \circ g)(1) = f(2) = 4$.

4. Let f and g be defined as in Problem 1d. Then f is not one-to-one, g is one-to-one, and $f \circ g$ is one-to-one.

5. a. Let $f : \mathbf{Z} \to \mathbf{Z}$ and $g : \mathbf{Z} \to \mathbf{Z}$ be defined by

$$f(x) = x, \quad g(x) = \begin{cases} \frac{x}{2} & \text{if } x \text{ is even} \\ x & \text{if } x \text{ is odd.} \end{cases}$$

The mapping f is one-to-one and the mapping g is onto, but the composition $f \circ g = g$ is not one-to-one, since $(f \circ g)(1) = (f \circ g)(2)$ and $1 \neq 2$.

b. Let $f : \mathbf{Z} \to \mathbf{Z}$ and $g : \mathbf{Z} \to \mathbf{Z}$ be defined by $f(x) = x^3$ and $g(x) = x$. The mapping f is one-to-one, the mapping g is onto, but the mapping $f \circ g$ given by $(f \circ g)(x) = x^3$ is not onto since there is no $x \in \mathbf{Z}$ such that $(f \circ g)(x) = 2$.

6. a. Let $f : \mathbf{Z} \to \mathbf{Z}$ and $g : \mathbf{Z} \to \mathbf{Z}$ be defined by

$$f(x) = \begin{cases} \frac{x}{2} & \text{if } x \text{ is even} \\ x & \text{if } x \text{ is odd} \end{cases} \qquad g(x) = x.$$

The mapping f is onto and the mapping g is one-to-one, but the composition $f \circ g = f$ is not one-to-one, since $(f \circ g)(1) = (f \circ g)(2)$ and $1 \neq 2$.

b. Let $f : \mathbf{Z} \to \mathbf{Z}$ and $g : \mathbf{Z} \to \mathbf{Z}$ be defined by $f(x) = x$ and $g(x) = x^3$. The mapping f is onto, the mapping g is one-to-one, but the mapping $f \circ g$ given by $(f \circ g)(x) = x^3$ is not onto since there is no $x \in \mathbf{Z}$ such that $(f \circ g)(x) = 2$.

8. a. Let $f(x) = x, g(x) = x^2$, and $h(x) = |x|$, for all $x \in \mathbf{Z}$.

b. Let $f(x) = x^2, g(x) = x$ and $h(x) = -x$, for all $x \in \mathbf{Z}$.

12. To prove that f is one-to-one, suppose $f(a_1) = f(a_2)$, for a_1 and a_2 in A. Since $g \circ f$ is onto, there exist α_1 and α_2 in A such that

$$a_1 = (g \circ f)(\alpha_1) \quad \text{and} \quad a_2 = (g \circ f)(\alpha_2).$$

Then $f((g \circ f)(\alpha_1)) = f((g \circ f)(\alpha_2))$, since $f(a_1) = f(a_2)$, or

$$(f \circ g)(f(\alpha_1)) = (f \circ g)(f(\alpha_2)).$$

This implies that

$$f(\alpha_1) = f(\alpha_2)$$

since $f \circ g$ is one-to-one. Since g is a mapping, then

$$g(f(\alpha_1)) = g(f(\alpha_2)).$$

Thus

$$(g \circ f)(\alpha_1) = (g \circ f)(\alpha_2)$$

and

$$a_1 = a_2.$$

Therefore f is one-to-one.

To show that f is onto, let $b \in B$. Then $g(b) \in A$ and therefore $g(b) = (g \circ f)(a)$ for some $a \in A$ since $g \circ f$ is onto. It follows then that

$$(f \circ g)(b) = (f \circ g)(f(a)).$$

Since $f \circ g$ is one-to-one, we have

$$b = f(a),$$

and f is onto.

Exercises 1.4

1. **a**. The set B is not closed, since $-1 \in B$ and $-1 * -1 = 1 \notin B$.
   b. The set B is not closed since $1 \in B$ and $2 \in B$ but $1 * 2 = 1 - 2 = -1 \notin B$.
   c. The set B is closed.
   d. The set B is closed.
   e. The set B is not closed, since $1 \in B$ and $1 * 1 = 0 \notin B$.
   f. The set B is closed.

2. **a**. Not commutative, Not associative, No identity element
   b. Not commutative, Associative, No identity element
   c. Not commutative, Not associative, No identity element
   d. Commutative, Not associative, No identity element
   e. Not commutative, Not associative, No identity element
   f. Commutative, Associative, 0 is an identity element. 0 is the only invertible element and its inverse is 0.
   g. Commutative, Associative, -3 is an identity element. $-x - 6$ is the inverse of x.
   h. Commutative, Not associative, No identity element
   i. Not commutative, Not associative, No identity element
   j. Commutative, Not associative, No identity element

3. **a**. The binary operation $*$ is not commutative since $B * C \neq C * B$.
   b. There is no identity element.

4. **a**. The binary operation $*$ is not commutative, since $D * A \neq A * D$.
   b. C is an identity element.

c. The elements A and B are inverses of each other and C is its own inverse.

Exercises 1.5

1. a. A right inverse does not exist since f is not onto.
   b. A right inverse does not exist since f is not onto.
   c. A right inverse $g : \mathbf{Z} \to \mathbf{Z}$ is defined by $g(x) = x - 2$.
   d. A right inverse $g : \mathbf{Z} \to \mathbf{Z}$ is defined by $g(x) = 1 - x$.
   e. A right inverse does not exist since f is not onto.
   f. A right inverse does not exist since f is not onto.
   g. A right inverse does not exist since f is not onto.
   h. A right inverse does not exist since f is not onto.
   i. A right inverse does not exist since f is not onto.
   j. A right inverse does not exist since f is not onto.
   k. A right inverse $g : \mathbf{Z} \to \mathbf{Z}$ is defined by $g(x) = \begin{cases} x & \text{if } x \text{ is even} \\ 2x + 1 & \text{if } x \text{ is odd.} \end{cases}$
   l. A right inverse does not exist since f is not onto.
   m. A right inverse $g : \mathbf{Z} \to \mathbf{Z}$ is defined by $g(x) = \begin{cases} 2x & \text{if } x \text{ is even} \\ x - 2 & \text{if } x \text{ is odd.} \end{cases}$
   n. A right inverse $g : \mathbf{Z} \to \mathbf{Z}$ is defined by $g(x) = \begin{cases} 2x - 1 & \text{if } x \text{ is even} \\ x - 1 & \text{if } x \text{ is odd.} \end{cases}$

2. a. A left inverse $g : \mathbf{Z} \to \mathbf{Z}$ is defined by $g(x) = \begin{cases} \frac{x}{2} & \text{if } x \text{ is even} \\ 1 & \text{if } x \text{ is odd.} \end{cases}$
   b. A left inverse $g : \mathbf{Z} \to \mathbf{Z}$ is defined by $g(x) = \begin{cases} \frac{x}{3} & \text{if } x \text{ is a multiple of 3} \\ 0 & \text{if } x \text{ is not a multiple of 3.} \end{cases}$
   c. A left inverse $g : \mathbf{Z} \to \mathbf{Z}$ is defined by $g(x) = x - 2$.
   d. A left inverse $g : \mathbf{Z} \to \mathbf{Z}$ is defined by $g(x) = 1 - x$.
   e. A left inverse $g : \mathbf{Z} \to \mathbf{Z}$ is defined by $g(x) = \begin{cases} y & \text{if } x = y^3 \text{ for some } y \in \mathbf{Z} \\ 0 & \text{if } x \neq y^3 \text{ for some } y \in \mathbf{Z}. \end{cases}$
   f. A left inverse does not exist since f is not one-to-one.
   g. A left inverse $g : \mathbf{Z} \to \mathbf{Z}$ is defined by $g(x) = \begin{cases} x & \text{if } x \text{ is even} \\ \dfrac{x+1}{2} & \text{if } x \text{ is odd.} \end{cases}$

h. A left inverse does not exist since f is not one-to-one.

i. A left inverse does not exist since f is not one-to-one.

j. A left inverse does not exist since f is not one-to-one.

k. A left inverse does not exist since f is not one-to-one.

l. A left inverse $g : \mathbf{Z} \to \mathbf{Z}$ is defined by: $g(x) = \begin{cases} x+1 & \text{if } x \text{ is odd} \\ \frac{x}{2} & \text{if } x \text{ is even} \end{cases}$

m. A left inverse does not exist since f is not one-to-one.

n. A left inverse does not exist since f is not one-to-one.

3. $n!$

4. Let $f : A \to A$, where A is nonempty.

$$\begin{aligned} f \text{ has a left inverse } &\Leftrightarrow f \text{ is one-to-one, by Lemma 1.23} \\ &\Leftrightarrow f^{-1}(f(S)) = S \text{ for every subset } S \text{ of } A, \text{ by} \\ &\qquad \text{Exercise 24 of Section 1.2.} \end{aligned}$$

5. Let $f : A \to A$, where A is nonempty.

$$\begin{aligned} f \text{ has a right inverse } &\Leftrightarrow f \text{ is onto, by Lemma 1.24} \\ &\Leftrightarrow f\left(f^{-1}(T)\right) = T \text{ for every subset } T \text{ of } A, \text{ by} \\ &\qquad \text{Exercise 25 of Section 1.2.} \end{aligned}$$

Exercises 1.6

1. **a**. $A = \begin{bmatrix} 1 & 0 \\ 3 & 2 \\ 5 & 4 \end{bmatrix}$ **b**. $A = \begin{bmatrix} -1 & -2 \\ 1 & 2 \\ -1 & -2 \\ 1 & 2 \end{bmatrix}$ **c**. $B = \begin{bmatrix} 1 & -1 & 1 & -1 \\ -1 & 1 & -1 & 1 \end{bmatrix}$

d. $B = \begin{bmatrix} 0 & 1 & 1 & 1 \\ 0 & 0 & 1 & 1 \\ 0 & 0 & 0 & 1 \end{bmatrix}$ **e**. $C = \begin{bmatrix} 2 & 0 & 0 \\ 3 & 4 & 0 \\ 4 & 5 & 6 \\ 5 & 6 & 7 \end{bmatrix}$ **f**. $C = \begin{bmatrix} 1 & 0 & 0 \\ 0 & 1 & 0 \\ 0 & 0 & 1 \\ 0 & 0 & 0 \end{bmatrix}$

2. **a**. $\begin{bmatrix} 3 & 0 & -4 \\ 8 & -8 & 6 \end{bmatrix}$ **b**. $\begin{bmatrix} 1 & 9 \\ -3 & 2 \end{bmatrix}$ **c**. Not possible **d**. Not possible

3. **a.** $\begin{bmatrix} -5 & 7 \\ 8 & -1 \end{bmatrix}$ **b.** $\begin{bmatrix} -10 & 2 & 1 \\ -14 & 6 & -21 \\ 6 & -1 & -2 \end{bmatrix}$ **c.** Not possible **d.** $\begin{bmatrix} 7 & -11 \\ 12 & 6 \\ -2 & 20 \end{bmatrix}$

e. $\begin{bmatrix} 4 & 2 \\ 3 & 7 \end{bmatrix}$ **f.** $\begin{bmatrix} 1 & 3 \\ -4 & 10 \end{bmatrix}$ **g.** Not possible **h.** Not possible

i. $[4]$ **j.** $\begin{bmatrix} -12 & 8 & -4 \\ -15 & 10 & -5 \\ 18 & -12 & 6 \end{bmatrix}$

4.
$$\begin{aligned} c_{ij} &= \sum_{k=1}^{3} (i+k)(2k-j) \\ &= (i+1)(2-j) + (i+2)(4-j) + (i+3)(6-j) \\ &= 12i - 6j - 3ij + 28 \end{aligned}$$

6.
$$\begin{bmatrix} 1 & 6 & -3 & 2 \\ 4 & -7 & 1 & 5 \end{bmatrix} \begin{bmatrix} w \\ x \\ y \\ z \end{bmatrix} = \begin{bmatrix} 9 \\ 0 \end{bmatrix}$$

7. **a.** n **b.** $n(n-1)$ **c.** 12
d. δ_{ik}, if $1 \le i \le n, 1 \le k \le n; 0$ if $i > n$ or $k > n$

8.

$\cdot$	I	A	B	C
I	I	A	B	C
A	A	B	C	I
B	B	C	I	A
C	C	I	A	B

9. (Answer not unique) $A = \begin{bmatrix} 1 & 2 \\ 3 & 4 \end{bmatrix}, B = \begin{bmatrix} 1 & 1 \\ 1 & 1 \end{bmatrix}$

10. A trivial example is with $A = I_2$ and B an arbitrary 2×2 matrix. Another example is provided by $A = \begin{bmatrix} 1 & 1 \\ 1 & 1 \end{bmatrix}$ and $B = \begin{bmatrix} 2 & 3 \\ 3 & 2 \end{bmatrix}$.

11. (Answer not unique) $A = \begin{bmatrix} 1 & 2 \\ 1 & 2 \end{bmatrix}, B = \begin{bmatrix} -6 & -6 \\ 3 & 3 \end{bmatrix}$

12. (Answer not unique) $A = \begin{bmatrix} 1 & 0 \\ 0 & 0 \end{bmatrix}, B = \begin{bmatrix} 2 & 3 \\ 4 & 5 \end{bmatrix}, C = \begin{bmatrix} 2 & 3 \\ 6 & 7 \end{bmatrix}$.

13. $(A-B)(A+B) = \begin{bmatrix} 10 & 1 \\ 2 & 1 \end{bmatrix}$ and $A^2 - B^2 = \begin{bmatrix} 2 & 6 \\ -4 & 9 \end{bmatrix}, (A-B)(A+B) \neq A^2 - B^2$.

14. $(A+B)^2 = \begin{bmatrix} 22 & 5 \\ 30 & 7 \end{bmatrix}, A^2+2AB+B^2 = \begin{bmatrix} 30 & 0 \\ 36 & -1 \end{bmatrix}, (A+B)^2 \neq A^2+2AB+B^2$.

15. $X = A^{-1}B$ **16**. $X = A^{-1}BC^{-1}$

20. **b**. For each x in G of the form $\begin{bmatrix} a & a \\ 0 & 0 \end{bmatrix}$, then $y = \begin{bmatrix} 1 & 1 \\ 0 & 0 \end{bmatrix}$. For each x in G of the form $\begin{bmatrix} 0 & 0 \\ a & a \end{bmatrix}$, then $y = \begin{bmatrix} 0 & 0 \\ 1 & 1 \end{bmatrix}$.

Exercises 1.7

1. **a**. This is a mapping since for every $a \in A$ there is a unique $b \in A$ such that (a, b) is an element of the relation.

   b. This is a mapping since for every $a \in A$ there is $1 \in A$ such that $(a, 1)$ is an element of the relation.

   c. This is not a mapping since the element 1 is related to three different values; $1R1$, $1R3$, and $1R5$.

   d. This is a mapping since for every $a \in A$ there is a unique $b \in A$ such that (a, b) is an element of the relation.

   e. This is a mapping since for every $a \in A$ there is a unique $b \in A$ such that (a, b) is an element of the relation.

   f. This is not a mapping since the element 5 is related to three different values: $5R1, 5R3$, and $5R5$.

2. **a**. The relation R is not reflexive since $2 \not R 2$. It is not symmetric since $4R2$ but $2 \not R 4$. It is not transitive since $4R2$ and $2R1$ but $4 \not R 1$.

   b. The relation R is not reflexive since $2 \not R 2$. It is symmetric since $x = -y \Rightarrow y = -x$. It is not transitive since $2R(-2)$ and $(-2)R2$, but $2 \not R 2$.

c. The relation R is reflexive and transitive, but not symmetric since for arbitrary x, y, and z in $\mathbf{Z}$ we have:

(1) $x = x \cdot 1$ with $1 \in \mathbf{Z}$

(2) $6 = 3\,(2)$ with $2 \in \mathbf{Z}$ but $3 \neq 6k$ where $k \in \mathbf{Z}$

(3) $y = xk_1$ for some $k_1 \in \mathbf{Z}$ and $z = yk_2$ for some $k_2 \in \mathbf{Z}$ imply $z = yk_2 = x\,(k_1 k_2)$ with $k_1 k_2 \in \mathbf{Z}$.

d. The relation R is not reflexive since $1 \not{R} 1$. It is not symmetric since $1R2$, but $2\not{R}1$. It is transitive since $x < y$ and $y < z \Rightarrow x < z$ for all x, y and $z \in \mathbf{Z}$.

e. The relation R is reflexive since $x \geq x$ for all $x \in \mathbf{Z}$. It is not symmetric since $5R3$ but $3\not{R}5$. It is transitive since $x \geq y$ and $y \geq z$ imply $x \geq z$ for all x, y, z in $\mathbf{Z}$.

f. The relation R is not reflexive since $(-1)\not{R}(-1)$. It is not symmetric since $1R(-1)$ but $(-1)\not{R}1$. It is transitive since $x = |y|$ and $y = |z|$ implies $x = |y| = ||z|| = |z|$ for all x, y and $z \in \mathbf{Z}$.

g. The relation R is not reflexive since $(-6)\not{R}(-6)$. It is not symmetric since $3R5$ but $5\not{R}3$. It is not transitive since $4R3$ and $3R2$, but $4\not{R}2$.

h. The relation R is reflexive since $x^2 \geq 0$ for all x in $\mathbf{Z}$. It is also symmetric since $xy \geq 0$ implies that $yx \geq 0$. It is not transitive since $(-2)\,R0$ and $0R4$ but $(-2)\not{R}4$.

i. The relation R is not reflexive since $2\not{R}2$. It is symmetric since $xy \leq 0$ implies $yx \leq 0$ for all $x, y \in \mathbf{Z}$. It is not transitive since $-1R2$ and $2R(-3)$ but $(-1)\not{R}(-3)$.

j. The relation R is not reflexive since $|x - x| = 0 \neq 1$. It is symmetric since $|x - y| = 1 \Rightarrow |y - x| = 1$. It is not transitive since $|2 - 1| = 1$ and $|1 - 2| = 1$ but $|2 - 2| = 0 \neq 1$.

k. The relation R is reflexive, symmetric and transitive since for arbitrary x, y and z in $\mathbf{Z}$, we have:

(1) $|x - x| = |0| < 1$

(2) $|x - y| < 1 \Rightarrow |y - x| < 1$

(3) $|x - y| < 1$ and $|y - z| < 1 \Rightarrow x = y$ and $y = z \Rightarrow |x - z| < 1$.

3. a. $\{-3, 3\}$ b. $\{-5, -1, 3, 7, 11\} \subseteq [3]$

4. b. $[0] = \{\ldots, -10, -5, 0, 5, 10, \ldots\}$, $[1] = \{\ldots, -9, -4, 1, 6, 11, \ldots\}$, $[2] = \{\ldots, -8, -3, 2, 7, 12, \ldots\}$, $[8] = [3] = \{\ldots, -7, -2, 3, 8, 13, \ldots\}$ $[-4] = [1] = \{\ldots, -9, -4, 1, 6, 11, \ldots\}$

5. b. $[0] = \{\ldots, -14, -7, 0, 7, 14, \ldots\}$, $[1] = \{\ldots, -13, -6, 1, 8, 15, \ldots\}$ $[3] = \{\ldots, -11, -4, 3, 10, 17, \ldots\}$, $[9] = [2] = \{\ldots, -12, -5, 2, 9, 16, \ldots\}$ $[-2] = [5] = \{\ldots, -9, -2, 5, 12, 19, \ldots\}$

6. $[0] = \{\ldots, -2, 0, 2, 4, \ldots\}$, $[1] = \{\ldots, -3, -1, 1, 3, \ldots\}$

7. $[0] = \{0, \pm 5, \pm 10, \ldots\}$ $[1] = \{\pm 1, \pm 4, \pm 6, \pm 9, \ldots\}$, $[2] = \{\pm 2, \pm 3, \pm 7, \pm 8, \ldots\}$

8. $[0] = \{\ldots, -4, 0, 4, 8, \ldots\}$, $[1] = \{\ldots, -7, -3, 1, 5, \ldots\}$,
$[2] = \{\ldots, -6, -2, 2, 6, \ldots\}$, $[3] = \{\ldots, -5, -1, 3, 7, \ldots\}$,

9. $[0] = \{\ldots, -7, 0, 7, 14, \ldots\}$, $[1] = \{\ldots, -13, -6, 1, 8, \ldots\}$,
$[2] = \{\ldots, -12, -5, 2, 9, \ldots\}$, $[3] = \{\ldots, -11, -4, 3, 10, \ldots\}$,
$[4] = \{\ldots, -10, -3, 4, 11, \ldots\}$, $[5] = \{\ldots, -9, -2, 5, 12, \ldots\}$,
$[6] = \{\ldots, -8, -1, 6, 13, \ldots\}$

10. a. The relation R is reflexive and transitive but not symmetric, since for arbitrary nonempty subsets x, y and z of A we have:

(1) x is a subset of x;
(2) x is a subset of y does not imply that y is a subset of x;
(3) x is a subset of y and y is a subset of z imply that x is a subset of z.

b. The relation R is not reflexive and not symmetric, but it is transitive since for arbitrary nonempty subsets x, y, and z of A we have:

(1) x is not a proper subset of x;
(2) x is a proper subset of y implies that y is not a proper subset of x;
(3) x is a proper subset of y and y is a proper subset of z imply that x is a proper subset of z.

c. The relation R is reflexive, symmetric and transitive since for arbitrary nonempty subsets x, y, and z of A we have:

(1) x and x have the same number of elements;
(2) If x and y have the same number of elements, then y and x have the same number of elements;
(3) If x and y have the same number of elements and y and z have the same number of elements, then x and z have the same number of elements.

11. a. The relation is reflexive and symmetric but not transitive since if x, y, and z are human beings, we have:

(1) x lives within 400 miles of x;
(2) x lives within 400 miles of y implies that y lives within 400 miles of x;
(3) x lives within 400 miles of y and y lives within 400 miles of z do not imply that x lives within 400 miles of z.

b. The relation R is not reflexive, not symmetric, and not transitive since if x, y, and z are human beings we have:

(1) x is not the father of x;
(2) x is the father of y implies that y is not the father of x;
(3) x is the father of y and y is the father of z imply that x is not the father of z.

c. The relation is symmetric but not reflexive and not transitive. Let x, y, and z be human beings, and we have:

(1) x is a first cousin of x is not a true statement;

(2) x is a first cousin of y implies that y is a first cousin of x;

(3) x is a first cousin of y and y is a first cousin of z do not imply that x is a first cousin of z.

d. The relation R is reflexive, symmetric, and transitive since if x, y, and z are human beings we have:

(1) x and x were born in the same year;

(2) if x and y were born in the same year, then y and x were born in the same year;

(3) if x and y were born in the same year and if y and z were born in the same year, then x and z were born in the same year.

e. The relation is reflexive, symmetric, and transitive since if x, y, and z are human beings, we have:

(1) x and x have the same mother;

(2) x and y have the same mother implies y and x have the same mother;

(3) x and y have the same mother and y and z have the same mother imply that x and z have the same mother.

12. a. The relation R is an equivalence relation on $A \times A$. Let a, b, c, d, p, and q be arbitrary elements of A.

(1) $(a, b)\, R\, (a, b)$ since $ab = ba$.

(2) $(a, b)\, R\, (c, d) \Rightarrow ad = bc \Rightarrow (c, d)\, R\, (a, b)$.

(3) $$\begin{aligned}(a, b)\, R\, (c, d) \text{ and } (c, d)\, R\, (p, q) &\Rightarrow ad = bc \text{ and } cq = dp \\ &\Rightarrow adcq = bcdp \\ &\Rightarrow aq = bp \text{ since } c \neq 0 \text{ and } d \neq 0 \\ &\Rightarrow (a, b)\, R\, (p, q).\end{aligned}$$

b. The relation R is an equivalence relation on $A \times A$. Let $(a, b), (c, d), (e, f)$ be arbitrary elements of $A \times A$.

(1) $(a, b)\, R\, (a, b)$ since $ab = ab$;

(2) $(a, b)\, R\, (c, d) \Rightarrow ab = cd \Rightarrow cd = ab \Rightarrow (c, d)\, R\, (a, b)$;

(3) $(a, b)\, R\, (c, d)$ and $(c, d)\, R\, (e, f) \Rightarrow ab = cd$ and $cd = ef \Rightarrow ab = ef \Rightarrow (a, b)\, R\, (e, f)$.

c. The relation R is an equivalence relation on $A \times A$. Let a, b, c, d, p, and q be arbitrary elements of A.

(1) $(a, b)\, R\, (a, b)$ since $a^2 + b^2 = a^2 + b^2$.

(2) $(a, b)\, R\, (c, d) \Rightarrow a^2 + b^2 = c^2 + d^2 \Rightarrow c^2 + d^2 = a^2 + b^2 \Rightarrow (c, d)\, R\, (a, b)$.

(3) $(a,b)\,R\,(c,d)$ and $(c,d)\,R\,(p,q)$ $\Rightarrow a^2+b^2=c^2+d^2$ and

$$c^2+d^2=p^2+q^2$$

$$\Rightarrow a^2+b^2=p^2+q^2$$

$$\Rightarrow (a,b)\,R\,(p,q).$$

d. The relation R is an equivalence relation on $A\times A$. Let (a,b), (c,d) and (e,f) be arbitrary elements of $A\times A$.

(1) $(a,b)\,R\,(a,b)$ since $a-b=a-b$;
(2) $(a,b)\,R\,(c,d)\Rightarrow a-b=c-d\Rightarrow c-d=a-b\Rightarrow (c,d)\,R\,(a,b)$;
(3) $(a,b)\,R\,(c,d)$ and $(c,d)\,R\,(e,f)\Rightarrow a-b=c-d$ and $c-d=e-f\Rightarrow a-b=e-f\Rightarrow (a,b)\,R\,(e,f)$.

13. The relation R is reflexive and symmetric but not transitive.

14. **a**. The relation is symmetric but not reflexive and not transitive. Let x, y, and z be arbitrary elements of the power set $\mathcal{P}(A)$ of the nonempty set A.

(1) $x\cap x\neq\varnothing$ is not true if $x=\varnothing$.
(2) $x\cap y\neq\varnothing$ implies that $y\cap x\neq\varnothing$.
(3) $x\cap y\neq\varnothing$ and $y\cap z\neq\varnothing$ do not imply that $x\cap z\neq\varnothing$. For example, let $A=\{a,b,c,d\}$, $x=\{b,c\}$, $y=\{c,d\}$ and $z=\{d,a\}$. Then $x\cap y=\{c\}\neq\varnothing, y\cap z=\{d\}\neq\varnothing$ but $x\cap z=\varnothing$.

b. The relation R is reflexive and transitive but not symmetric since for arbitrary subsets x,y,z of A we have:

(1) $x\subseteq x$;
(2) $\varnothing\subseteq A$ but $A\nsubseteq\varnothing$;
(3) $x\subseteq y$ and $y\subseteq z$ imply $x\subseteq z$.

15. The relation is reflexive, symmetric, and transitive. Let x,y, and z be arbitrary elements of the power set $\mathcal{P}(A)$ and C a fixed subset of A.

(1) xRx since $x\cap C=x\cap C$.

(2) $xRy\Rightarrow x\cap C=y\cap C\Rightarrow y\cap C=x\cap C\Rightarrow yRx$.

(3) xRy and yRz $\Rightarrow x\cap C=y\cap C$ and $y\cap C=z\cap C$

$$\Rightarrow x\cap C=z\cap C$$

$$\Rightarrow xRz.$$

Thus R is an equivalence relation on $\mathcal{P}(A)$.

16. **a**. The relation R is reflexive, symmetric, and transitive. Let a,b, and c represent arbitrary triangles in the plane. Then

(1) a is similar to a is true;

(2) a is similar to b implies that b is similar to a;

(3) a is similar to b and b is similar to c imply that a is similar to c.

b. The relation R is reflexive, symmetric, and transitive. Let a, b, and c represent arbitrary triangles in the plane. Then

(1) a is congruent to a is true;

(2) a is congruent to b implies that b is congruent to a;

(3) a is congruent to b and b is congruent to c imply that a is congruent to c.

17. d, j **18**. d **19**. a, d, e, f, k

21. $\bigcup_{\lambda\in\mathcal{L}} A_\lambda = A_1 \cup A_2 \cup A_3 = \{a, b, c, d, e, f, g\}$, $\bigcap_{\lambda\in\mathcal{L}} A_\lambda = A_1 \cap A_2 \cap A_3 = \{c\}$

22. $\bigcup_{\lambda\in\mathcal{L}} A_\lambda = \mathbf{Z}$, $\bigcap_{\lambda\in\mathcal{L}} A_\lambda = \varnothing$

24. The only property of an equivalence relation that fails to hold is the transitive property: $1R2$ and $2R3$, but $1\not R 3$.

Exercises 2.1

17. Hint: Use Exercise 14 twice to obtain $0 < xz < yz$ and $0 < yz < yw$. The transitive property (Exercise 13) then yields $xz < yw$.

23. Hint: consider the following cases:

(1) $0 \le x$ and $0 \le y$

(2) $0 \le x, y < 0$ and $0 \le x + y$

(3) $0 \le x, y < 0$ and $x + y < 0$

(4) $x < 0$ and $0 \le y$ is covered by cases (2) and (3) because of the symmetry in x and y.

Outline of proof for case (2):

$$0 \le x,\ y < 0, \text{ and } 0 \le x + y \Rightarrow |x| = x, |y| = -y, \text{ and } |x + y| = x + y$$

$$\left.\begin{aligned} y < 0 \Rightarrow 0 < -y \Rightarrow x < x - y \\ y < 0 \Rightarrow x + y < x \end{aligned}\right\} \Rightarrow x + y < x - y$$

$$\Rightarrow |x + y| < |x| + |y|.$$

24. Outline of proof:

$$\begin{aligned} a \text{ positive and } b \text{ negative } &\Rightarrow 0 < a \text{ and } b < 0 \\ &\Rightarrow a \in \mathbf{Z}^{+} \text{ and } 0 - b = -b \in \mathbf{Z}^{+} \\ &\Rightarrow a(-b) = -ab \in \mathbf{Z}^{+} \\ &\Rightarrow ab < 0 \\ &\Rightarrow ab \text{ is negative} \end{aligned}$$

25. Hint: Assume b is negative and obtain an immediate contradiction to Exercise 24.

27. All the addition postulates and all the multiplication postulates except 2c are satisfied. Postulate 2c is not satisfied since $\{0\}$ does not contain an element different from 0. The set $\{0\}$ has the properties required in postulate 4, and postulate 5 is satisfied vacuously (that is, there is no counterexample). Thus, all postulates except 2c are satisfied.

Exercises 2.2

26. Now $0 < 1 \Rightarrow 0 + 1 < 1 + 1$, by Exercise 11 of Section 2.1. Thus $1 < 2 = 2^1$, and the statement is true for $n = 1$.

Assume now that $k < 2^k$ for the positive integer k. By the transitive properties of $\leq$ and $<$,

$$1 \leq k \text{ and } k < 2^k \Rightarrow 1 < 2^k.$$

Now

$$\begin{aligned} k < 2^k \text{ and } 1 < 2^k &\Rightarrow k + 1 < 2^k + 2^k \text{ by Exercise 12, Section 2.1} \\ &\Rightarrow k + 1 < 2^k(1 + 1) \\ &\Rightarrow k + 1 < 2^k(2) = 2^{k+1} \\ &\Rightarrow k + 1 < 2^{k+1} \end{aligned}$$

Thus the statement is true for $n = k + 1$, and therefore true for all positive integers.

Exercises 2.3

1. **a**. $\pm1, \pm2, \pm3, \pm5, \pm6, \pm10, \pm15, \pm30$
b. $\pm1, \pm2, \pm3, \pm6, \pm7, \pm14, \pm21, \pm42$
c. $\pm1, \pm2, \pm4, \pm7, \pm14, \pm28$ **d**. $\pm1, \pm3, \pm5, \pm9, \pm15, \pm45$
e. $\pm1, \pm2, \pm3, \pm4, \pm6, \pm8, \pm12, \pm24$
f. $\pm1, \pm2, \pm4, \pm5, \pm8, \pm10, \pm20, \pm40$ **g**. $\pm1, \pm2, \pm4, \pm8, \pm16, \pm32$

h. ± 1, ± 2, ± 3, ± 5, ± 6, ± 7, ± 10, ± 14, ± 15, ± 21, ± 30, ± 35, ± 42, ± 70, ± 105, ± 210

2. a. $\pm 1, \pm 2$ b. $\pm 1, \pm 3$ c. $\pm 1, \pm 2, \pm 4, \pm 8$ d. $\pm 1, \pm 2, \pm 5, \pm 10$
e. $\pm 1, \pm 2, \pm 4, \pm 8$ f. $\pm 1, \pm 2, \pm 5, \pm 10$

3. $q = 30, r = 16$ **4**. $q = 34, r = 2$ **5**. $q = 22, r = 5$ **6**. $q = 32, r = 21$

7. $q = -3, r = 3$ **8**. $q = -4, r = 1$ **9**. $q = -51, r = 4$ **10**. $q = -52, r = 15$

11. $q = 0, r = 26$ **12**. $q = 0, r = 15$ **13**. $q = -360, r = 3$

14. $q = -32, r = 156$ **15**. $q = 0, r = 0$ **16**. $q = 0, r = 0$

23. Counterexample: Let $a = 6, b = 8$, and $c = 9$.

26. Let m be an arbitrary positive integer and assume there is an integer n such that $m < n < m+1$. Then n is a positive integer, since $0 < m$ and $m < n$. By Exercise 11 of Section 2.1,

$$m < n < m + 1 \Rightarrow 0 < n - m < 1.$$

Thus $n - m$ is a positive integer less than 1, and this contradicts Theorem 2.6. Therefore the assumption that there is a positive integer n such that $m < n < m+1$ is false, and the proof is complete.

27. If $a = 0$, then $n = -1$ makes $a - bn = 0 - b(-1) = b > 0$, and we have a positive element of S in this case. If $a \neq 0$, the choice $n = -2|a|$ gives $a - bn = a + 2b|a|$ as a specific example of a positive element of S. The problem does not explicitly require a proof that our element is positive, but this can be done as follows.

Since $b > 0$, we have $b \geq 1$ by Theorem 2.6. This implies $b|a| \geq |a|$ by Exercise 14 of Section 2.1. It follows from the definition of absolute value that $|a| \geq -a$. Now

$$b|a| \geq |a| \quad \text{and} \quad |a| \geq -a \Rightarrow b|a| \geq -a.$$

Since $a \neq 0, |a| > 0$, and therefore $|a| \geq 1$ by Theorem 2.6. Hence, $b|a| \geq b$ by Exercise 14 of Section 2.1.

$$b|a| \geq b \quad \text{and} \quad b > 0 \Rightarrow b|a| > 0$$

We have $b|a| \geq -a$ and $b|a| > 0$. By Exercise 12 of Section 2.1,

$$\begin{aligned} b|a| + b|a| &> -a + 0, \\ 2b|a| &> -a, \text{ and} \\ a + 2b|a| &> 0. \end{aligned}$$

This shows that $a + 2b|a|$ is positive.

Exercises 2.4

1. 2, 3, 5, 7, 11, 13, 17, 19, 23, 29, 31, 37, 41, 43, 47, 53, 59, 61, 67, 71, 73, 79, 83, 89, 97

2. a. $1400 = 2^3 \cdot 5^2 \cdot 7$, $980 = 2^2 \cdot 5 \cdot 7^2$, $(1400, 980) = 2^2 \cdot 5 \cdot 7 = 140$

   b. $4950 = 2 \cdot 3^2 \cdot 5^2 \cdot 11$, $10,500 = 2^2 \cdot 3 \cdot 5^3 \cdot 7$, $(4950,\ 10,500) = 2 \cdot 3 \cdot 5^2 = 150$

   c. $3780 = 2^2 \cdot 3^3 \cdot 5 \cdot 7$, $16,200 = 2^3 \cdot 3^4 \cdot 5^2$, $(3780,\ 16,200) = 2^2 \cdot 3^3 \cdot 5 = 540$

   d. $52,920 = 2^3 \cdot 3^3 \cdot 5 \cdot 7^2$, $25,200 = 2^4 \cdot 3^2 \cdot 5^2 \cdot 7$,
   $(52,920,\ 25,200) = 2^3 \cdot 3^2 \cdot 5 \cdot 7 = 2520$

3. a. $(a, b) = 3, m = 0, n = -1$ b. $(a, b) = 13, m = 3, n = 2$

   c. $(a, b) = 6, m = 2, n = -3$ d. $(a, b) = 4, m = 12, n = -5$

   e. $(a, b) = 3, m = 2, n = 25$ f. $(a, b) = 36, m = -2, n = -3$

   g. $(a, b) = 9, m = -5, n = 3$ h. $(a, b) = 2, m = 3, n = -44$

   i. $(a, b) = 3, m = -49, n = 188$ j. $(a, b) = 60, m = 7, n = -22$

   k. $(a, b) = 12, m = -3, n = 146$ l. $(a, b) = 14, m = -5, n = 12$

   m. $(a, b) = 12, m = 5, n = 163$ n. $(a, b) = 1, m = 19, n = -661$

4. a. $(4, 6) = 2$ b. $(6, 15) = 3$

6.

n	1	2	3	4	5
$n^2 - n + 5$	5	7	11	17	25 (not prime)

Show that $n^2 - n + 11$ is a prime integer when $n = 1, 2, ..., 10$, but that it is not true that $n^2 - n + 11$ is always a prime integer.

22. After a and b are written in their standard forms, the least common multiple of a and b can be found by forming the product of all the distinct prime factors that appear in the standard form of either a or b, with each factor raised to the greatest power to which it appears in either standard form.

23. a. $1400 = 2^3 \cdot 5^2 \cdot 7$, $980 = 2^2 \cdot 5 \cdot 7^2$; the least common multiple is $2^3 \cdot 5^2 \cdot 7^2 = 9800$

   b. $4950 = 2 \cdot 3^2 \cdot 5^2 \cdot 11$, $10,500 = 2^2 \cdot 3 \cdot 5^3 \cdot 7$; the least common multiple is $2^2 \cdot 3^2 \cdot 5^3 \cdot 7 \cdot 11 = 346,500$

   c. $3780 = 2^2 \cdot 3^3 \cdot 5 \cdot 7$, $16,200 = 2^3 \cdot 3^4 \cdot 5^2$; the least common multiple is $2^3 \cdot 3^4 \cdot 5^2 \cdot 7 = 113,400$

   d. $52,920 = 2^3 \cdot 3^3 \cdot 5 \cdot 7^2$, $25,200 = 2^4 \cdot 3^2 \cdot 5^2 \cdot 7$; the least common multiple is $2^4 \cdot 3^3 \cdot 5^2 \cdot 7^2 = 529,200$

24. a. An integer d is a **greatest common divisor** of a, b, and c if these conditions are satisfied:

(1) d is a positive integer;
(2) $d \mid a$, $d \mid b$, and $d \mid c$;
(3) If $n \mid a$, $n \mid b$, and $n \mid c$, then $n \mid d$.

Exercises 2.5

1. $[0] = \{\ldots, -5, 0, 5, \ldots\}$, $[1] = \{\ldots, -4, 1, 6, \ldots\}$, $[2] = \{\ldots, -3, 2, 7, \ldots\}$
$[3] = \{\ldots, -2, 3, 8, \ldots\}$, $[4] = \{\ldots, -1, 4, 9, \ldots\}$

2. $[0] = \{..., -6, 0, 6, ...\}$, $[1] = \{..., -5, 1, 7, ...\}$, $[2] = \{..., -4, 2, 8, ...\}$
$[3] = \{..., -3, 3, 9, ...\}$, $[4] = \{..., -2, 4, 10, ...\}$, $[5] = \{..., -1, 5, 11, ...\}$

3. $x = 5$ **4**. $x = 4$ **5**. $x = 11$ **6**. $x = 10$ **7**. $x = 8$ **8**. $x = 7$

9. $x = 173$ **10**. $x = 63$ **11**. $x = 28$ **12**. $x = 39$ **13**. $x = 7$

14. $x = 2$ **15**. $x = 4$ **16**. $x = 12$ **17**. $x = 6$ **18**. $x = 216$

19. $x = 11$ **20**. $x = 8$ **21**. $x = 13$ **22**. $x = 31$ **23**. $x = 2$

24. $x = 37$ **29**. **a**. 3 **b**. 1

37. $d = (6, 27) = 3$ and 3 divides 33; $x = 1, x = 10$, and $x = 19$ are solutions

38. $d = (18, 15) = 3$, and 3 divides 33; $x = 1, x = 6$, and $x = 11$ are solutions.

39. $d = (8, 78) = 2$, and 2 divides 66; $x = 18$ and $x = 57$ are solutions.

40. $d = (35, 20) = 5$, and 5 divides 10; $x = 2, x = 6, x = 10, x = 14$, and $x = 18$ are solutions.

41. $d = (68, 40) = 4$, and 4 divides 36; $x = 7, x = 17, x = 27$, and $x = 37$ are solutions.

42. $d = (21, 30) = 3$, and 3 divides 18; $x = 8, x = 18$, and $x = 28$ are solutions.

43. $d = (24, 348) = 12$, and 12 does not divide 45; therefore, there are no solutions.

44. $d = (36, 270) = 18$, and 18 does not divide 48; therefore there are no solutions.

45. $d = (15, 110) = 5$, and 5 divides 130; $x = 16, x = 38, x = 60, x = 82$, and $x = 104$ are solutions.

46. $d = (20, 76) = 4$, and 4 divides 124; $x = 10, x = 29, x = 48$, and $x = 67$ are solutions.

47. $d = (42, 74) = 2$, and 2 divides 30; $x = 6$ and $x = 43$ are solutions.

48. $d = (38, 60) = 2$, and 2 divides 26; $x = 7$ and $x = 37$ are solutions.

52. **a**. $x = 27$ or $x \equiv 27 \pmod{40}$ **b**. $x = 14$ or $x \equiv 14 \pmod{15}$

Exercises 2.6

1. **a**. [3] **b**. [7] **c**. [4] **d**. [6] **e**. [6] [4] = [0] **f**. [11]
   g. [6] + [6] = [0] **h**. [11]

2. **a**. [1] [2] [3] [4] = [24] = [4] **b**. [1] [2] [3] [4] [5] [6] = [720] = [6]
   c. [1] [2] [3] = [6] = [2] **d**. [0]

3. **a**.

+	[0]	[1]
[0]	[0]	[1]
[1]	[1]	[0]

b.

+	[0]	[1]	[2]
[0]	[0]	[1]	[2]
[1]	[1]	[2]	[0]
[2]	[2]	[0]	[1]

c.

+	[0]	[1]	[2]	[3]	[4]
[0]	[0]	[1]	[2]	[3]	[4]
[1]	[1]	[2]	[3]	[4]	[0]
[2]	[2]	[3]	[4]	[0]	[1]
[3]	[3]	[4]	[0]	[1]	[2]
[4]	[4]	[0]	[1]	[2]	[3]

d.

+	[0]	[1]	[2]	[3]	[4]	[5]
[0]	[0]	[1]	[2]	[3]	[4]	[5]
[1]	[1]	[2]	[3]	[4]	[5]	[0]
[2]	[2]	[3]	[4]	[5]	[0]	[1]
[3]	[3]	[4]	[5]	[0]	[1]	[2]
[4]	[4]	[5]	[0]	[1]	[2]	[3]
[5]	[5]	[0]	[1]	[2]	[3]	[4]

e.

+	[0]	[1]	[2]	[3]	[4]	[5]	[6]
[0]	[0]	[1]	[2]	[3]	[4]	[5]	[6]
[1]	[1]	[2]	[3]	[4]	[5]	[6]	[0]
[2]	[2]	[3]	[4]	[5]	[6]	[0]	[1]
[3]	[3]	[4]	[5]	[6]	[0]	[1]	[2]
[4]	[4]	[5]	[6]	[0]	[1]	[2]	[3]
[5]	[5]	[6]	[0]	[1]	[2]	[3]	[4]
[6]	[6]	[0]	[1]	[2]	[3]	[4]	[5]

f.

+	[0]	[1]	[2]	[3]	[4]	[5]	[6]	[7]
[0]	[0]	[1]	[2]	[3]	[4]	[5]	[6]	[7]
[1]	[1]	[2]	[3]	[4]	[5]	[6]	[7]	[0]
[2]	[2]	[3]	[4]	[5]	[6]	[7]	[0]	[1]
[3]	[3]	[4]	[5]	[6]	[7]	[0]	[1]	[2]
[4]	[4]	[5]	[6]	[7]	[0]	[1]	[2]	[3]
[5]	[5]	[6]	[7]	[0]	[1]	[2]	[3]	[4]
[6]	[6]	[7]	[0]	[1]	[2]	[3]	[4]	[5]
[7]	[7]	[0]	[1]	[2]	[3]	[4]	[5]	[6]

4. a.

×	[0]	[1]
[0]	[0]	[0]
[1]	[0]	[1]

b.

×	[0]	[1]	[2]
[0]	[0]	[0]	[0]
[1]	[0]	[1]	[2]
[2]	[0]	[2]	[1]

c.

×	[0]	[1]	[2]	[3]	[4]	[5]
[0]	[0]	[0]	[0]	[0]	[0]	[0]
[1]	[0]	[1]	[2]	[3]	[4]	[5]
[2]	[0]	[2]	[4]	[0]	[2]	[4]
[3]	[0]	[3]	[0]	[3]	[0]	[3]
[4]	[0]	[4]	[2]	[0]	[4]	[2]
[5]	[0]	[5]	[4]	[3]	[2]	[1]

d.

×	[0]	[1]	[2]	[3]	[4]
[0]	[0]	[0]	[0]	[0]	[0]
[1]	[0]	[1]	[2]	[3]	[4]
[2]	[0]	[2]	[4]	[1]	[3]
[3]	[0]	[3]	[1]	[4]	[2]
[4]	[0]	[4]	[3]	[2]	[1]

e.

×	[0]	[1]	[2]	[3]	[4]	[5]	[6]
[0]	[0]	[0]	[0]	[0]	[0]	[0]	[0]
[1]	[0]	[1]	[2]	[3]	[4]	[5]	[6]
[2]	[0]	[2]	[4]	[6]	[1]	[3]	[5]
[3]	[0]	[3]	[6]	[2]	[5]	[1]	[4]
[4]	[0]	[4]	[1]	[5]	[2]	[6]	[3]
[5]	[0]	[5]	[3]	[1]	[6]	[4]	[2]
[6]	[0]	[6]	[5]	[4]	[3]	[2]	[1]

f.

×	[0]	[1]	[2]	[3]	[4]	[5]	[6]	[7]
[0]	[0]	[0]	[0]	[0]	[0]	[0]	[0]	[0]
[1]	[0]	[1]	[2]	[3]	[4]	[5]	[6]	[7]
[2]	[0]	[2]	[4]	[6]	[0]	[2]	[4]	[6]
[3]	[0]	[3]	[6]	[1]	[4]	[7]	[2]	[5]
[4]	[0]	[4]	[0]	[4]	[0]	[4]	[0]	[4]
[5]	[0]	[5]	[2]	[7]	[4]	[1]	[6]	[3]
[6]	[0]	[6]	[4]	[2]	[0]	[6]	[4]	[2]
[7]	[0]	[7]	[6]	[5]	[4]	[3]	[2]	[1]

5. **a**. [9] **b**. [8] **c**. [13] **d**. [22] **e**. [5] **f**. [51] **g**. [173]
h. [57]

6. **a**. [1] , [5] **b**. [1] , [3] , [5] , [7] **c**. [1] , [3] , [7] , [9] **d**. [1] , [5] , [7] , [11]

e. [1] , [5] , [7] , [11] , [13] , [17] **f**. [1] , [3] , [7] , [9] , [11] , [13] , [17] , [19]

7. **a**. [2] , [3] , [4] **b**. [2] , [4] , [6] **c**. [2] , [4] , [5] , [6] , [8]
d. [2] , [3] , [4] , [6] , [8] , [9] , [10]

e. [2] , [3] , [4] , [6] , [8] , [9] , [10] , [12] , [14] , [15] , [16]
f. [2] , [4] , [5] , [6] , [8] , [10] , [12] , [14] , [15] , [16] , [18]

8. **a**. $[x] = [2]$ or $[x] = [5]$ **b**. No solution exists. **c**. $[x] = [2]$ or $[x] = [6]$
d. $[x] = [3]$ or $[x] = [9]$ **e**. No solution exists. **f**. No solution exists.

g. $[x]=[2]$, $[x]=[5]$, $[x]=[8]$, or $[x]=[11]$ h. $[x]=[6]$ or $[x]=[13]$
i. $[x]=[4]$ or $[x]=[10]$ j. $[x]=[3]$, $[x]=[7]$, or $[x]=[11]$

10. a. $[x]=[4]^{-1}[5]=[10][5]=[11]$ b. $[x]=[8]^{-1}[7]=[7][7]=[5]$
c. $[x]=[7]^{-1}[11]=[7][11]=[5]$ d. $[x]=[8]^{-1}[11]=[2][11]=[7]$
e. $[x]=[9]^{-1}[14]=[9][14]=[6]$ f. $[x]=[8]^{-1}[15]=[17][15]=[12]$
g. $[x]=[6]^{-1}[5]=[266][5]=[54]$ h. $[x]=[9]^{-1}[8]=[27][8]=[216]$

11. $[x]=[3],[y]=[5]$ **12.** $[x]=[3],[y]=[5]$ **13.** $[x]=[3],[y]=[3]$

14. $[x]=[1],[y]=[5]$

19. a. $[x]=[4]$ or $[x]=[5]$ b. $[x]=[4]$ or $[x]=[2]$ c. $[x]=[1]$ or $[x]=[5]$
d. $[x]=[1]$ or $[x]=[3]$

Exercises 2.7

1. Errors occur in 00010 and 11100

2. Errors occur in 111011, 011110 and 001000

3. Corrected coded message:

101101101 110110110 100100100 101101101 010010010 011011011

Decoded message: 101 110 100 101 010 011

4. Corrected coded message:

11110 01011 01011 10101 01011 10101 11110

Decoded message: 11 01 01 10 01 10 11

5. a. $\frac{3}{4}$ b. $\frac{3}{6}=\frac{1}{2}$ c. $\frac{2}{6}=\frac{1}{3}$ d. $\frac{4}{9}$

6. a. $(0.97)^4+4(0.97)^3(0.03)=0.9948136$ b. $(0.03)^4=0.00000081$

7. a. $(0.9999)^8=0.9992003$ b. $8(0.9999)^7(0.0001)=0.0007994402$
c. $(0.9999)^8+8(0.9999)^7(0.0001)=0.9999997$
d. $\binom{8}{2}(0.9999)^6(0.0001)^2=0.0000002798320$ e. 1.000000

8. $(0.999)^{400}=0.6701859$ **9.** 1 **10.** No, 0 is the check digit.

14. a. 7 b. 4 c. 1 d. 7

17. a. Valid b. Not valid c. Not valid d. Not valid

18. **a**. No error is detected. **b**. An error is detected. **c**. An error is detected.
d. An error is detected.

19. $\mathbf{y} = -(10, 9, 8, 7, 6, 5, 4, 3, 2)$ **20**. **a**. 3 **b**. 3 **c**. 3 **d**. 5

22. **a**. 3 **b**. 4 **c**. 3 **d**. 3 **23**. 2 **24**. 3 **25**. 3

26. 00000 01010 10011 11001 00101 01111 10110 11100

Exercises 2.8

1. Ciphertext: APMHKPMKSHQ HQVHAPMHUIQT
$f^{-1}(x) = x + 19 \bmod 27$

2. Ciphertext: M.VJTIQKFRQYEXU
$f^{-1}(x) = x + 10 \bmod 31$

3. Plaintext: "tiger, do you read me?"
$f^{-1}(x) = x + 20 \bmod 31$

4. Plaintext: "rise and shine"
$f^{-1}(x) = x + 12 \bmod 31$

5. Ciphertext: FBBZXLXDGIXZUW
$f^{-1}(x) = 4x + 7 \bmod 27$

6. Ciphertext: DPMN.PAYJLUJDW,UJWJ?'PGBUQI
$f^{-1}(x) = 29x + 13 \bmod 31$

7. Plaintext: www.brookscole.com
$f^{-1}(x) = 19x + 2 \bmod 28$

8. Plaintext: www.uscs.edu
$f^{-1}(x) = 23x + 7 \bmod 28$

9. Plaintext: mathematics
$f(x) = 9x + 13 \bmod 26$, $f^{-1}(x) = 3x + 13 \bmod 26$

10. Plaintext: one small step for man
$f(x) = 2x + 5 \bmod 27$, $f^{-1}(x) = 14x + 11 \bmod 27$

11. Plaintext: there are 25 primes less than 100
$f(x) = 12x + 17 \bmod 37$, $f^{-1}(x) = 34x + 14 \bmod 37$

12. Plaintext: been there, done that.
$f(x) = 8x + 22 \bmod 29$
$f^{-1}(x) = 11x + 19 \bmod 29$

15. **a**. $n - 1$ **b**. $(n-1)n - 1 = n^2 - n - 1$

16. Ciphertext: 455 617 416 189 272 617 001 551 446 320

17. Ciphertext: 62 49 75 26 49 73 75 50 61 $d = 37$

18. Ciphertext: 20 00 19 26 38 49 26 06 00 24 49 23 $d = 11$

19. a. Ciphertext: 000 132 085 082 001 030 000

b. Ciphertext: 001 050 105 039 000 c. $d = 103$

20. a. Ciphertext: 281 000 018 059 000 277

b. Ciphertext: 245 018 248 277 c. $d = 91$

21. Plaintext: quaternions **22**. Plaintext: euclidean algorithm

23. a. $\phi(5) = 4;\quad 1, 2, 3, 4$

b. $\phi(19) = 18;\quad 1, 2, 3, 4, 5, 6, 7, 8, 9, 10, 11, 12, 13, 14, 15, 16, 17, 18$

c. $\phi(15) = 8;\quad 1, 2, 4, 7, 8, 11, 13, 14$

d. $\phi(27) = 18;\quad 1, 2, 4, 5, 7, 8, 10, 11, 13, 14, 16, 17, 19, 20, 22, 23, 25, 26$

e. $\phi(12) = 4;\quad 1, 5, 7, 11$

f. $\phi(36) = 12;\quad 1, 5, 7, 11, 13, 17, 19, 23, 25, 29, 31, 35$

24. The positive integers less than or equal to pq that are not relatively prime to pq are either multiples of p, that is, elements of the set

$$P = \{1p,\ 2p,\ 3p,\ \dots,\ (q-1)\,p,\ qp\},$$

or multiples of q, that is, elements of the set

$$Q = \{1q,\ 2q,\ 3q,\ \dots,\ (p-1)\,q,\ pq\}.$$

Since there are q elements in P, and p elements in Q, and the element pq has been counted twice, we have

$$\phi(pq) = pq - q - p + 1 = (p-1)(q-1).$$

25. The positive integers less than or equal to p^j that are not relatively prime to p^j are multiples of p, that is, elements of the set

$$\{1p, 2p, 3p, \dots (p^{j-1} - 1)\,p, p^{j-1}p\}.$$

Since this set contains p^{j-1} elements, then

$$\phi(p^j) = p^j - p^{j-1} = p^{j-1}(p-1).$$

Exercises 3.1

1. Group

2. The set of all irrational numbers with the operation of addition does not form a group because there is no identity element, and the set is not closed under addition.

3. The set of all positive irrational numbers with the operation of multiplication does not form a group. The set is not closed with respect to multiplication. For example, $\sqrt{2}$ is a positive irrational number, but $\sqrt{2}\sqrt{2} = 2$ is not. Also, there is no identity element.

4. Group

5. The set of all real numbers x such that $0 < x \leq 1$ is not a group with respect to multiplication because not all elements have inverses.

6. Group

7. Group

8. The set of all complex numbers x that have absolute value 1 with the operation of addition does not form a group. The set is not closed with respect to addition. (The complex number 1 has absolute value 1, and the complex number i has absolute value 1, but the sum $1 + i$ has absolute value $\sqrt{2}$.) There is no identity element in the set.

9. Group

10. The set $\mathbf{E}$ of all even integers with the operation of multiplication does not form a group because there is no identity element, and the set does not contain inverses.

11. The operation $\times$ is not associative, since

$$a \times (c \times a) = a \times e = a,$$

whereas

$$(a \times c) \times a = b \times a = c.$$

Also, there are no inverses for the elements a and b.

12. There is no identity element.

13.

$\times$	a	b	c	d
a	c	d	a	b
b	d	c	b	a
c	a	b	c	d
d	b	a	d	c

14.

$\times$	a	b	c	d
a	c	d	a	b
b	d	a	b	c
c	a	b	c	d
d	b	c	d	a

15. The set **Z** is an abelian group with respect to $*$. The identity element is -1. The element $-x-2$ is the inverse of the element $x \in \mathbf{Z}$.

16. The set **Z** is an abelian group with respect to $*$. The identity element is 1. The element $2-x$ is the inverse of the element $x \in \mathbf{Z}$.

17. The set **Z** is not a group and hence not an abelian group with respect to the operation $*$. The operation is not associative. There is no identity element and hence no inverse elements.

18. The set **Z** is not a group and hence not an abelian group with respect to $*$. The operation $*$ is not associative. There is no identity element and hence no inverse elements.

19. The set **Z** is not a group and hence not an abelian group with respect to $*$. The identity element is 0, but 1 does not have an inverse in **Z**.

20. The set **Z** is not a group and hence not an abelian group with respect to $*$. The operation $*$ is not associative. There is no identity element and hence no inverse elements.

21. Group, 2 **22**. Group, 4

23. The set is not a group with respect to multiplication, since it does not have an identity element and hence no inverse elements.

24. The set is not a group with respect to multiplication. The identity element is $[6]$, but the element $[0]$ has no inverse.

25. Group, 5 **26**. Group, 4

27. a. $n-1$ b.

×	[1]	[2]	[3]	[4]	[5]	[6]
[1]	[1]	[2]	[3]	[4]	[5]	[6]
[2]	[2]	[4]	[6]	[1]	[3]	[5]
[3]	[3]	[6]	[2]	[5]	[1]	[4]
[4]	[4]	[1]	[5]	[2]	[6]	[3]
[5]	[5]	[3]	[1]	[6]	[4]	[2]
[6]	[6]	[5]	[4]	[3]	[2]	[1]

$[1]^{-1} = [1]$
$[2]^{-1} = [4]$
$[3]^{-1} = [5]$
$[4]^{-1} = [2]$
$[5]^{-1} = [3]$
$[6]^{-1} = [6]$

28.

$\cdot$	1	i	j	k	-1	$-i$	$-j$	$-k$
1	1	i	j	k	-1	$-i$	$-j$	$-k$
i	i	-1	k	$-j$	$-i$	1	$-k$	j
j	j	$-k$	-1	i	$-j$	k	1	$-i$
k	k	j	$-i$	-1	$-k$	$-j$	i	1
-1	-1	$-i$	$-j$	$-k$	1	i	j	k
$-i$	$-i$	1	$-k$	j	i	-1	k	$-j$
$-j$	$-j$	k	1	$-i$	j	$-k$	-1	i
$-k$	$-k$	$-j$	i	1	k	j	$-i$	-1

29.

$\times$	I_3	P_1	P_2	P_3	P_4	P_5
I_3	I_3	P_1	P_2	P_3	P_4	P_5
P_1	P_1	I_3	P_3	P_2	P_5	P_4
P_2	P_2	P_5	I_3	P_4	P_3	P_1
P_3	P_3	P_4	P_1	P_5	P_2	I_3
P_4	P_4	P_3	P_5	P_1	I_3	P_2
P_5	P_5	P_2	P_4	I_3	P_1	P_3

30.

$\times$	I_2	R	R^2	R^3	H	D	V	T
I_2	I_2	R	R^2	R^3	H	D	V	T
R	R	R^2	R^3	I_2	D	V	T	H
R^2	R^2	R^3	I_2	R	V	T	H	D
R^3	R^3	I_2	R	R^2	T	H	D	V
H	H	T	V	D	I_2	R^3	R^2	R
D	D	H	T	V	R	I_2	R^3	R^2
V	V	D	H	T	R^2	R	I_2	R^3
T	T	V	D	H	R^3	R^2	R	I_2

34. The set G is not a group with respect to addition since it does not contain an identity element with respect to addition.

35. The set G is not a group with respect to addition since it does not contain an identity element with respect to addition.

36. **a.** $\begin{bmatrix} [4] & [2] & [5] \\ [0] & [1] & [3] \end{bmatrix}$ **b.** $\begin{bmatrix} [5] & [3] & [6] \\ [0] & [2] & [4] \end{bmatrix}$ **37.** **a.** $\begin{bmatrix} [3] & [1] \\ [4] & [2] \end{bmatrix}$ **b.** $\begin{bmatrix} [5] & [1] \\ [5] & [3] \end{bmatrix}$

42. One possible choice is $a = \rho$ and $b = \sigma$. Then $(ab)^{-1} = (\rho \circ \sigma)^{-1} = \gamma^{-1} = \gamma$ and $a^{-1}b^{-1} = \rho^{-1} \circ \sigma^{-1} = \rho^2 \circ \sigma = \delta$, so $(ab)^{-1} \neq a^{-1}b^{-1}$.

43. One possible choice for the elements, a, b, and c is the following: $a = \rho, b = \sigma$, and $c = \rho^2$. Then we have $\rho \circ \sigma = \gamma = \sigma \circ \rho^2$, but $\rho \neq \rho^2$.

44. One possible choice is $a = \rho$ and $b = \delta$. Then $(ab)^2 = (\rho \circ \delta)^2 = \sigma^2 = e$ and $a^2b^2 = \rho^2 \circ \delta^2 = \rho^2 \circ e = \rho^2$, so $(ab)^2 \neq a^2b^2$.

50. $(abcd)^{-1} = \left(\left(d^{-1}c^{-1}\right) b^{-1}\right) a^{-1}$

52. Consider the set $S = \left\{a \in G \mid a \neq a^{-1}\right\}$. Now $a \in S$ if and only if $a^{-1} \in S$, so S has an even number of elements. Since both G and S have an even number of elements, the complement $G - S = \left\{a \in G \mid a = a^{-1}\right\}$ must also have an even number of elements. The element e is in $G - S$ since $e = e^{-1}$ and therefore there is at least one $a \neq e$ such that $a = a^{-1}$.

53. **b.** 2^n

54. $\mathcal{P}(A) = \{\varnothing, \{a\}, \{b\}, \{c\}, \{a,b\}, \{a,c\}, \{b,c\}, A\}$

$+$	$\varnothing$	$\{a\}$	$\{b\}$	$\{c\}$	$\{a,b\}$	$\{a,c\}$	$\{b,c\}$	A
$\varnothing$	$\varnothing$	$\{a\}$	$\{b\}$	$\{c\}$	$\{a,b\}$	$\{a,c\}$	$\{b,c\}$	A
$\{a\}$	$\{a\}$	$\varnothing$	$\{a,b\}$	$\{a,c\}$	$\{b\}$	$\{c\}$	A	$\{b,c\}$
$\{b\}$	$\{b\}$	$\{a,b\}$	$\varnothing$	$\{b,c\}$	$\{a\}$	A	$\{c\}$	$\{a,c\}$
$\{c\}$	$\{c\}$	$\{a,c\}$	$\{b,c\}$	$\varnothing$	A	$\{a\}$	$\{b\}$	$\{a,b\}$
$\{a,b\}$	$\{a,b\}$	$\{b\}$	$\{a\}$	A	$\varnothing$	$\{b,c\}$	$\{a,c\}$	$\{c\}$
$\{a,c\}$	$\{a,c\}$	$\{c\}$	A	$\{a\}$	$\{b,c\}$	$\varnothing$	$\{a,b\}$	$\{b\}$
$\{b,c\}$	$\{b,c\}$	A	$\{c\}$	$\{b\}$	$\{a,c\}$	$\{a,b\}$	$\varnothing$	$\{a\}$
A	A	$\{b,c\}$	$\{a,c\}$	$\{a,b\}$	$\{c\}$	$\{b\}$	$\{a\}$	$\varnothing$

55. The set $\varnothing$ is an identity element, but the set $\mathcal{P}(A)$ does not contain inverse elements. Hence $\mathcal{P}(A)$ is not a group with respect to the operation of union.

56. The set A is an identity element. But the set $\mathcal{P}(A)$ is not a group with respect to the operation of intersection, since A is the only element that has an inverse.

58. Let G be a set with a binary operation $*$ on G satisfying the associative property and conditions 1 and 2.

By condition 1, there exists a left identity e such that $e * x = x$ for all x in G. We shall show that $x * e = x$ also holds for all $x \in G$.

Let x be a fixed element of G. By condition 2, there exists x' in G such that $x' * x = e$, and also there exists y in G such that $y * x' = e$. Now

$$y * (x' * x) = y * e \text{ since } x' * x = e$$

and

$$\begin{aligned}(y * x') * x &= e * x \quad \text{since } y * x' = e \\ &= x \qquad \text{since } e \text{ is a left identity.}\end{aligned}$$

By the associative property, $x = y * e$. But

$$\begin{aligned} x = y * e \Rightarrow x * e &= (y * e) * e \\ &= y * (e * e) \\ &= y * e \qquad \text{since } e \text{ is a left identity} \\ &= x \qquad \text{since } x = y * e. \end{aligned}$$

Thus e is an identity element for $*$ on G.

But we have $x = y * e$ and this means that $x = y$ and therefore

$$\begin{aligned} x * x' &= y * x' \\ &= e \qquad \text{since } y * x' = e. \end{aligned}$$

Hence x' is an inverse of x because $x' * x = e$ and $x * x' = e$. This proves that G is a group.

59. Let n be a positive integer, $n \geq 2$. For elements $a_1, a_2, \ldots, a_n$ in a group G, the expression $a_1 + a_2 + \cdots + a_n$ is defined recursively by

$$a_1 + a_2 + \cdots + a_k + a_{k+1} = (a_1 + a_2 + \cdots + a_k) + a_{k+1}, \quad \text{for } k \geq 1.$$

Exercises 3.2

1. **a**. The set $\{e, \sigma\}$ is a subgroup of $\mathcal{S}(A)$. The multiplication table is:

$\circ$	e	σ
e	e	σ
σ	σ	e

b. The set $\{e, \delta\}$ is a subgroup of $\mathcal{S}(A)$. The multiplication table is:

$\circ$	e	δ
e	e	δ
δ	δ	e

c. The set $\{e, \rho\}$ is not a subgroup of $\mathcal{S}(A)$ since it is not closed. We have $\rho \circ \rho = \rho^2 \notin \{e, \rho\}$. The multiplication table is:

$\circ$	e	ρ
e	e	ρ
ρ	ρ	ρ^2

d. The set $\{e, \rho^2\}$ is not a subgroup of $\mathcal{S}(A)$ since it is not closed. We have $\rho^2 \circ \rho^2 = \rho \notin \{e, \rho^2\}$. The multiplication table is:

$\circ$	e	ρ^2
e	e	ρ^2
ρ^2	ρ^2	ρ

e. The set $\{e, \rho, \rho^2\}$ is a subgroup of $\mathcal{S}(A)$. The multiplication table is:

$\circ$	e	ρ	ρ^2
e	e	ρ	ρ^2
ρ	ρ	ρ^2	e
ρ^2	ρ^2	e	ρ

f. The set $\{e, \rho, \sigma\}$ is not a subgroup of $\mathcal{S}(A)$ since it is not closed. We have $\rho \circ \rho = \rho^2 \notin \{e, \rho, \sigma\}$. The multiplication table is:

$\circ$	e	ρ	σ
e	e	ρ	σ
ρ	ρ	ρ^2	γ
σ	σ	δ	e

g. The set $\{e, \sigma, \gamma\}$ is not a subgroup of $\mathcal{S}(A)$ since it is not closed. We have $\gamma \circ \sigma = \rho \notin \{e, \sigma, \gamma\}$. The multiplication table is:

$\circ$	e	σ	γ
e	e	σ	γ
σ	σ	e	ρ^2
γ	γ	ρ	e

h. The set $\{e, \sigma, \gamma, \delta\}$ is not a subgroup of $\mathcal{S}(A)$ since it is not closed. We have $\sigma \circ \gamma = \rho^2 \notin \{e, \sigma, \gamma, \delta\}$. The multiplication table is:

$\circ$	e	σ	γ	δ
e	e	σ	γ	δ
σ	σ	e	ρ^2	ρ
γ	γ	ρ	e	ρ^2
δ	δ	ρ^2	ρ	e

2. **a**. Subgroup

b. The set $\{1, i\}$ is not a subgroup of the group $G = \{1, -1, i, -i\}$ under multiplication because $\{1, i\}$ is not closed. We have $i \cdot i = -1 \notin \{1, i\}$.

c. The set $\{i, -i\}$ is not a subgroup of G, since it is not closed. We have $i \cdot i = -1 \notin \{i, -i\}$.

d. The set $\{1, -i\}$ is not a subgroup of the group $G = \{1, -1, i, -i\}$ under multiplication because $\{1, -i\}$ is not closed. We have $(-i) \cdot (-i) = -1 \notin \{1, -i\}$.

3. $\langle [6] \rangle = \{[0], [2], [4], [6], [8], [10], [12], [14]\}, \quad o(\langle [6] \rangle) = 8$

4. $\langle [8] \rangle = \{[0], [2], [4], [6], [8], [10], [12], [14], [16]\}, \quad o(\langle [8] \rangle) = 9$

5. **a**. $\{[1], [3], [4], [9], [10], [12]\}, \quad o(\langle [4] \rangle) = 6$

b. $\{[1], [5], [8], [12]\}, \quad o(\langle [8] \rangle) = 4$

6. **a**. $\langle A \rangle = \left\{ \begin{bmatrix} 0 & -1 \\ 1 & 0 \end{bmatrix}, \begin{bmatrix} -1 & 0 \\ 0 & -1 \end{bmatrix}, \begin{bmatrix} 0 & 1 \\ -1 & 0 \end{bmatrix}, \begin{bmatrix} 1 & 0 \\ 0 & 1 \end{bmatrix} \right\}, \quad o(\langle A \rangle) = 4$

b. $\langle A \rangle = \{A, I_2\}, \quad o(\langle A \rangle) = 2$

7. **a**. $\langle A \rangle = \left\{ \begin{bmatrix} [2] & [0] \\ [0] & [3] \end{bmatrix}, \begin{bmatrix} [4] & [0] \\ [0] & [1] \end{bmatrix}, \begin{bmatrix} [1] & [0] \\ [0] & [4] \end{bmatrix}, \begin{bmatrix} [3] & [0] \\ [0] & [2] \end{bmatrix}, \begin{bmatrix} [0] & [0] \\ [0] & [0] \end{bmatrix} \right\}$,

$o(\langle A \rangle) = 5$

b. $\langle A\rangle = \left\{ \begin{bmatrix} [0] & [1] \\ [2] & [4] \end{bmatrix}, \begin{bmatrix} [0] & [2] \\ [4] & [3] \end{bmatrix}, \begin{bmatrix} [0] & [3] \\ [1] & [2] \end{bmatrix}, \begin{bmatrix} [0] & [4] \\ [3] & [1] \end{bmatrix}, \begin{bmatrix} [0] & [0] \\ [0] & [0] \end{bmatrix} \right\}$,

$o(\langle A\rangle) = 5$

8. The set $\mathbf{Z}^+$ of positive integers is closed under addition but it does not contain inverses. Thus $\mathbf{Z}^+$ is not a subgroup of the additive group $\mathbf{Z}$.

9. The set of all real numbers that are greater than 1 is closed under multiplication but is not a subgroup of G, since it does not contain inverses. (If $x > 1$, then $x^{-1} < 1$.)

11. The set K is nonempty since $e \in H$ (H is a subgroup of G) and e can be written as

$$e = aea^{-1} \in K.$$

The set K is closed since if $x \in K$ and $y \in K$ then

$$x = ah_1a^{-1} \quad \text{for some } h_1 \in H$$

and

$$y = ah_2a^{-1} \quad \text{for some } h_2 \in H.$$

Then

$$\begin{aligned} xy &= \left(ah_1a^{-1}\right)\left(ah_2a^{-1}\right) \\ &= ah_1\left(a^{-1}a\right)h_2a^{-1} \\ &= ah_1h_2a^{-1} \\ &= aha^{-1} \end{aligned}$$

where $h = h_1h_2 \in H$ since H is closed.

The set K contains inverses since if $x \in K$ then

$$x = aha^{-1} \quad \text{for some } h \in H.$$

Now $h^{-1} \in H$, since H is a subgroup of G. We can write $x^{-1} = ah^{-1}a^{-1} \in K$ and

$$xx^{-1} = x^{-1}x = e.$$

18. a. $\{1, -1\}$ b. $\{I_2, R^2\}$ c. $\{I_3\}$ d. $\{kI_2, k \neq 0\}$

25. The subgroup $\langle m\rangle \cap \langle n\rangle$ is the set of all multiples of the least common multiple of m and n.

27. Let $H_1 = \{e, \sigma\}$ and $H_2 = \{e, \gamma\}$.

29. Let $H_1 = \{e, \sigma\}$ and $H_2 = \{e, \gamma\}$.

35. Assume H is a subgroup of G. Then H is closed.

Assume H is closed. To prove that H is a subgroup of G, we need only show that H contains inverses. Since G is a finite group, we may assume that H contains exactly n elements:

$$H = \{h_1, h_2, ..., h_n\}.$$

Consider the n products

$$h_1h_1,\ h_1h_2,\ h_1h_3,\ ...,\ h_1h_n.$$

Since G is a group, these n elements are distinct by Theorem 3.4e. Since H is closed

$$h_1h_j \in H \quad \text{for } j = 1, 2, ..., n.$$

Hence

$$h_1 = h_1h_i \qquad \text{for some } i.$$

Since G is a group,

$$h_1 = h_1e = h_1h_i$$

implies

$$e = h_i \quad \text{by Theorem 3.4e.}$$

Thus $e \in H$, and therefore $e = h_1h_j$ for some $h_j \in H$. Since G is a group,

$$e = h_1h_1^{-1}$$

and Theorem 3.4e implies

$$h_1^{-1} = h_j.$$

Thus H contains inverses, and hence H is a subgroup of G.

Exercises 3.3

1. $\langle e\rangle = \{e\}, \langle\rho\rangle = \{e, \rho, \rho^2\}, \langle\sigma\rangle = \{e, \sigma\}, \langle\gamma\rangle = \{e, \gamma\}, \langle\delta\rangle = \{e, \delta\}$

2. $\langle 1\rangle = \{1\}, \langle -1\rangle = \{\pm 1\}, \langle i\rangle = \{\pm 1, \pm i\}, \langle j\rangle = \{\pm 1, \pm j\}, \langle k\rangle = \{\pm 1, \pm k\}$

3. The element e has order 1. Each of the elements σ, γ, and δ has order 2. Each of the elements ρ and ρ^2 has order 3.

4. The element 1 has order one. The element -1 has order two. Each of the elements $\pm i, \pm j$, and $\pm k$ has order four.

5. $o(I_3) = 1, o(P_1) = o(P_2) = o(P_4) = 2, o(P_3) = o(P_5) = 3$

6. **a**. $o(A) = 2$ **b**. $o(A) = 4$

7. **a**. $[1], [3], [5], [7]$ **b**. $[1], [5], [7], [11]$ **c**. $[1], [3], [7], [9]$
d. $[1], [2], [4], [7], [8], [11], [13], [14]$ **e**. $[1], [3], [5], [7], [9], [11], [13], [15]$

f. $[1], [5], [7], [11], [13], [17]$

8. **a.** $\{[0]\}$, $\{[0], [6]\}$, $\{[0], [4], [8]\}$, $\{[0], [3], [6], [9]\}$, $\{[0], [2], [4], [6], [8], [10]\}$, $\mathbf{Z}_{12}$

b. $\{[0]\}$, $\{[0], [4]\}$, $\{[0], [2], [4], [6]\}$, $\mathbf{Z}_8$

c. $\{[0]\}$, $\{[0], [5]\}$, $\{[0], [2], [4], [6], [8]\}$, $\mathbf{Z}_{10}$

d. $\{[0]\}$, $\{[0], [5], [10]\}$, $\{[0], [3], [6], [9], [12]\}$, $\mathbf{Z}_{15}$

e. $\{[0]\}$, $\{[0], [8]\}$, $\{[0], [4], [8], [12]\}$, $\{[0], [2], [4], [6], [8], [10], [12], [14]\}$, $\mathbf{Z}_{16}$

f. $\{[0]\}$, $\{[0], [9]\}$, $\{[0], [6], [12]\}$, $\{[0], [3], [6], [9], [12], [15]\}$, $\{[0], [2], [4], [6], [8], [10], [12], [14], [16]\}$, $\mathbf{Z}_{18}$

9. **a.** $G = \langle [3] \rangle = \langle [5] \rangle$ **b.** $G = \langle [2] \rangle = \langle [3] \rangle$

c. $G = \langle [2] \rangle = \langle [6] \rangle = \langle [7] \rangle = \langle [8] \rangle$

d. $G = \langle [2] \rangle = \langle [6] \rangle = \langle [7] \rangle = \langle [11] \rangle$

e. $G = \langle [3] \rangle = \langle [5] \rangle = \langle [6] \rangle = \langle [7] \rangle = \langle [10] \rangle = \langle [11] \rangle = \langle [12] \rangle = \langle [14] \rangle$

f. $G = \langle [2] \rangle = \langle [3] \rangle = \langle [10] \rangle = \langle [13] \rangle = \langle [14] \rangle = \langle [15] \rangle$

10. **a.** $[3], [5]$ **b.** $[2], [3]$ **c.** $[2], [6], [7], [8]$ **d.** $[2], [6], [7], [11]$

e. $[3], [5], [6], [7], [10], [11], [12], [14]$

f. $[2], [3], [10], [13], [14], [15]$

11. **a.** $\{[1]\}$, $\{[1], [6]\}$, $\{[1], [2], [4]\}$, G **b.** $\{[1]\}$, $\{[1], [4]\}$, G

c. $\{[1]\}$, $\{[1], [10]\}$, $\{[1], [3], [4], [5], [9]\}$, G

d. $\{[1]\}$, $\{[1], [12]\}$, $\{[1], [3], [9]\}$, $\{[1], [5], [8], [12]\}$, $\{[1], [3], [4], [9], [10], [12]\}$, G

e. $\{[1]\}$, $\{[1], [16]\}$, $\{[1], [4], [13], [16]\}$, $\{[1], [2], [4], [8], [9], [13], [15], [16]\}$, G

f. $\{[1]\}$, $\{[1], [18]\}$, $\{[1], [7], [11]\}$, $\{[1], [7], [8], [11], [12], [18]\}$, $\{[1], [4], [5], [6], [7], [9], [11], [16], [17]\}$, G

13. **b.** $$H = \left\{ \begin{bmatrix} 1 & 0 \\ 0 & 1 \end{bmatrix}, \begin{bmatrix} 0 & -1 \\ 1 & 0 \end{bmatrix}, \begin{bmatrix} -1 & 0 \\ 0 & -1 \end{bmatrix}, \begin{bmatrix} 0 & 1 \\ -1 & 0 \end{bmatrix} \right\}$$

c. $$H = \left\{ \begin{bmatrix} 1 & 0 \\ 0 & 1 \end{bmatrix}, \begin{bmatrix} -\frac{1}{2} & -\frac{\sqrt{3}}{2} \\ \frac{\sqrt{3}}{2} & -\frac{1}{2} \end{bmatrix}, \begin{bmatrix} -\frac{1}{2} & \frac{\sqrt{3}}{2} \\ -\frac{\sqrt{3}}{2} & -\frac{1}{2} \end{bmatrix} \right\}$$

16. a. $G = \{[1], [3], [7], [9], [11], [13], [17], [19]\}$

$\cdot$	[1]	[3]	[7]	[9]	[11]	[13]	[17]	[19]
[1]	[1]	[3]	[7]	[9]	[11]	[13]	[17]	[19]
[3]	[3]	[9]	[1]	[7]	[13]	[19]	[11]	[17]
[7]	[7]	[1]	[9]	[3]	[17]	[11]	[19]	[13]
[9]	[9]	[7]	[3]	[1]	[19]	[17]	[13]	[11]
[11]	[11]	[13]	[17]	[19]	[1]	[3]	[7]	[9]
[13]	[13]	[19]	[11]	[17]	[3]	[9]	[1]	[7]
[17]	[17]	[11]	[19]	[13]	[7]	[1]	[9]	[3]
[19]	[19]	[17]	[13]	[11]	[9]	[7]	[3]	[1]

b. $G = \{[1], [3], [5], [7]\}$

$\cdot$	[1]	[3]	[5]	[7]
[1]	[1]	[3]	[5]	[7]
[3]	[3]	[1]	[7]	[5]
[5]	[5]	[7]	[1]	[3]
[7]	[7]	[5]	[3]	[1]

c. $G = \{[1], [5], [7], [11], [13], [17], [19], [23]\}$

$\cdot$	[1]	[5]	[7]	[11]	[13]	[17]	[19]	[23]
[1]	[1]	[5]	[7]	[11]	[13]	[17]	[19]	[23]
[5]	[5]	[1]	[11]	[7]	[17]	[13]	[23]	[19]
[7]	[7]	[11]	[1]	[5]	[19]	[23]	[13]	[17]
[11]	[11]	[7]	[5]	[1]	[23]	[19]	[17]	[13]
[13]	[13]	[17]	[19]	[23]	[1]	[5]	[7]	[11]
[17]	[17]	[13]	[23]	[19]	[5]	[1]	[11]	[7]
[19]	[19]	[23]	[13]	[17]	[7]	[11]	[1]	[5]
[23]	[23]	[19]	[17]	[13]	[11]	[7]	[5]	[1]

d. $G = \{[1], [7], [11], [13], [17], [19], [23], [29]\}$

$\cdot$	[1]	[7]	[11]	[13]	[17]	[19]	[23]	[29]
[1]	[1]	[7]	[11]	[13]	[17]	[19]	[23]	[29]
[7]	[7]	[19]	[17]	[1]	[29]	[13]	[11]	[23]
[11]	[11]	[17]	[1]	[23]	[7]	[29]	[13]	[19]
[13]	[13]	[1]	[23]	[19]	[11]	[7]	[29]	[17]
[17]	[17]	[29]	[7]	[11]	[19]	[23]	[1]	[13]
[19]	[19]	[13]	[29]	[7]	[23]	[1]	[17]	[11]
[23]	[23]	[11]	[13]	[29]	[1]	[17]	[19]	[7]
[29]	[29]	[23]	[19]	[17]	[13]	[11]	[7]	[1]

17. **a.** Not cyclic **b.** Not cyclic **c.** Not cyclic **d.** Not cyclic

18. $\langle[1]\rangle = \{[1]\}$, $\langle[8]\rangle = \{[1], [8]\}$, $\langle[4]\rangle = \langle[7]\rangle = \{[1], [4], [7]\}, \langle[2]\rangle = \langle[5]\rangle = G$

19. $a, a^5, a^7, a^{11}, a^{13}, a^{17}, a^{19}, a^{23}$

20. $\langle a\rangle = G$

$\langle a^2\rangle = \{a^2, a^4, a^6, a^8, a^{10}, a^{12}, a^{14}, a^{16}, a^{18}, a^{20}, a^{22}, a^{24} = e\}$

$\langle a^3\rangle = \{a^3, a^6, a^9, a^{12}, a^{15}, a^{18}, a^{21}, a^{24} = e\}$

$\langle a^4\rangle = \{a^4, a^8, a^{12}, a^{16}, a^{20}, a^{24} = e\}$

$\langle a^6\rangle = \{a^6, a^{12}, a^{18}, a^{24} = e\}$

$\langle a^8\rangle = \{a^8, a^{16}, a^{24} = e\}$

$\langle a^{12}\rangle = \{a^{12}, a^{24} = e\}$

$\langle a^{24}\rangle = \langle e\rangle = \{e\}$

21. All subgroups of $\mathbf{Z}$ are of the form $\langle n\rangle$, n a fixed integer.

22. The generators of the infinite cyclic group $\langle a\rangle$ are a and a^{-1}. **29.** $p - 1$

30. Let $o(a) = p, o(b) = q, o(ab) = k$, and let m be the least common multiple of p and q.

It follows from the definition of least common multiple (see Exercise 19 of Section 2.4) that $m = pr$ and $m = qs$ for r, s in $\mathbf{Z}$. Thus

$$\begin{aligned}
(ab)^m &= a^m b^m && \text{since } ab = ba \\
&= a^{pr} b^{qs} && \text{since } m = pr = qs \\
&= e^r \cdot e^s && \text{since } o(a) = p \text{ and } o(b) = q \\
&= e,
\end{aligned}$$

and this implies that $k \mid m$, by Exercise 26 of this section.

On the other hand,

$$\begin{aligned}
(ab)^k = e \Rightarrow a^k b^k &= e && \text{since } ab = ba \\
\Rightarrow a^k &= b^{-k} \\
\Rightarrow a^k &= e \text{ and } b^{-k} = e && \text{since } \langle a \rangle \cap \langle b \rangle = \{e\} \\
\Rightarrow a^k &= e \text{ and } b^k = e.
\end{aligned}$$

But $a^k = e$ implies $p \mid k$ and $b^k = e$ implies $q \mid k$, by Exercise 26 of this section. From the definition of the least common multiple (see Exercise 19 of Section 2.4), this implies that $m \mid k$.

We have shown that $k \mid m$ and that $m \mid k$. Therefore $m = k$.

31. Suppose a is an element of order m in the group G. Let k be an integer, and let $d = (k, m)$. We shall use Corollary 3.19 to show that a^k has order $\frac{m}{d}$.

By Exercise 17 of Section 2.4, $k = k_0 d$ and $m = m_0 d$, where $(k_0, m_0) = 1$. We note that $m_0 = \frac{m}{d}$ is a positive integer. We have

$$\begin{aligned}
\left(a^k\right)^{m_0} &= a^{k m_0} \\
&= a^{k_0 d m_0} && \text{since } k = k_0 d \\
&= \left(a^{m_0 d}\right)^{k_0} \\
&= \left(a^m\right)^{k_0} && \text{since } m = m_0 d \\
&= e^{k_0} \\
&= e.
\end{aligned}$$

Thus m_0 is a positive integer such that $\left(a^k\right)^{m_0} = e$.

To show that m_0 is the least such positive integer, let j be any positive integer such that $\left(a^k\right)^j = e$.

$$\begin{aligned}
a^{kj} = e &\Rightarrow m \mid kj && \text{by Exercise 26} \\
&\Rightarrow kj = mq && \text{for some } q \in \mathbf{Z} \\
&\Rightarrow k_0 dj = m_0 dq && \text{since } k = k_0 d, m = m_0 d \\
&\Rightarrow k_0 j = m_0 q \\
&\Rightarrow m_0 \mid k_0 j \\
&\Rightarrow m_0 \mid j && \text{since } (m_0, k_0) = 1.
\end{aligned}$$

Thus $m_0 \leq j$, and $m_0 = \frac{m}{d}$ is the order of a^k by Corollary 3.19.

Exercises 3.4

3. Let $\phi : \mathbf{Z}_4 \to G$ be defined by

$$\phi([0]_4) = [1]_5, \quad \phi([1]_4) = [2]_5, \quad \phi([2]_4) = [4]_5, \quad \phi([3]_4) = [3]_5.$$

4. Let $\phi : G \to G'$ be defined by

$$\phi(1) = [0], \quad \phi(i) = [3], \quad \phi(-1) = [2], \quad \phi(-i) = [1].$$

5. Let $\phi : H \to \mathcal{S}(A)$ be defined by

$$\phi(I_2) = I_A, \ \phi(M_1) = \sigma, \ \phi(M_2) = \rho, \ \phi(M_3) = \rho^2, \ \phi(M_4) = \gamma, \ \phi(M_5) = \delta.$$

6. Let $\phi : \mathbf{Z}_6 \to G$ be defined by

$$\phi([0]_6) = [1]_7, \quad \phi([1]_6) = [3]_7, \quad \phi([2]_6) = [2]_7,$$
$$\phi([3]_6) = [6]_7, \quad \phi([4]_6) = [4]_7, \quad \phi([5]_6) = [5]_7.$$

7. Let $\phi : \mathbf{Z} \to H$ be defined by $\phi(n) = \begin{bmatrix} 1 & n \\ 0 & 1 \end{bmatrix}, n \in \mathbf{Z}$. Then

$$\phi(n+m) = \begin{bmatrix} 1 & n+m \\ 0 & 1 \end{bmatrix} = \begin{bmatrix} 1 & n \\ 0 & 1 \end{bmatrix} \begin{bmatrix} 1 & m \\ 0 & 1 \end{bmatrix} = \phi(n) \cdot \phi(m)$$

for all $n, m \in \mathbf{Z}$.

8. Let $\phi : G \to H$ be defined by $\phi(1) = \begin{bmatrix} 1 & 0 \\ 0 & 1 \end{bmatrix}, \phi(i) = \begin{bmatrix} i & 0 \\ 0 & -i \end{bmatrix}$,

$$\phi(-1) = \begin{bmatrix} -1 & 0 \\ 0 & -1 \end{bmatrix}, \phi(-i) = \begin{bmatrix} -i & 0 \\ 0 & i \end{bmatrix}.$$

9. Define $\phi : G \to H$ by $\phi(a + bi) = \begin{bmatrix} a & -b \\ b & a \end{bmatrix}$ for $a + bi \in G$. Let $x = a + bi \in G$

and $y = c + di \in G$. Then

$$\begin{aligned}
\phi(xy) &= \phi((a+bi)(c+di)) \\
&= \phi((ac-bd)+(bc+ad)i) \\
&= \begin{bmatrix} ac-bd & -bc-ad \\ bc+ad & ac-bd \end{bmatrix} \\
&= \begin{bmatrix} a & -b \\ b & a \end{bmatrix} \begin{bmatrix} c & -d \\ d & c \end{bmatrix} \\
&= \phi(a+bi)\,\phi(c+di) \\
&= \phi(x)\,\phi(y).
\end{aligned}$$

13. If G is nonabelian, then ϕ is not an isomorphism. If a and b are elements of G such that $ab \neq ba$, then $(ab)^{-1} \neq (ba)^{-1}$ and therefore $\phi(ab) = (ab)^{-1} = b^{-1}a^{-1} \neq a^{-1}b^{-1} = \phi(a)\,\phi(b)$.

14. ϕ is an automorphism.

19. For notational convenience we let a represent $[a]$. The elements 3 and $3^5 = 5$ are generators of G. The automorphisms of G are ϕ_1 and ϕ_2 defined by:

$$\phi_1 : \begin{cases} \phi_1(1) = 1 \\ \phi_1(2) = 2 \\ \phi_1(3) = 3 \\ \phi_1(4) = 4 \\ \phi_1(5) = 5 \\ \phi_1(6) = 6 \end{cases} \quad \text{and} \quad \phi_2 : \begin{cases} \phi_2(3) = 5 \\ \phi_2(3^2) = \phi_2(2) = 5^2 = 4 \\ \phi_2(3^3) = \phi_2(6) = 5^3 = 6 \\ \phi_2(3^4) = \phi_2(4) = 5^4 = 2 \\ \phi_2(3^5) = \phi_2(5) = 5^5 = 3 \\ \phi_2(3^6) = \phi_2(1) = 5^6 = 1. \end{cases}$$

20. **a**. 2 **b**. 2 **c**. 4 **d**. 2

24. Let $\phi_1 : G \to H$ and $\phi_2 : G \to H$ be defined by:

$$\phi_1 : \begin{cases} \phi_1([1]) = [0] \\ \phi_1([2]) = [2] \\ \phi_1([3]) = [1] \\ \phi_1([4]) = [4] \\ \phi_1([5]) = [5] \\ \phi_1([6]) = [3] \end{cases} \quad \text{and} \quad \phi_2 : \begin{cases} \phi_2([1]) = [0] \\ \phi_2([2]) = [4] \\ \phi_2([3]) = [5] \\ \phi_2([4]) = [2] \\ \phi_2([5]) = [1] \\ \phi_2([6]) = [3]. \end{cases}$$

Exercises 3.5

1. a. ϕ is a homomorphism, $\ker \phi = \{1, -1\}$. ϕ is not an epimorphism because there is no x such that $\phi(x) = -1$.
   b. ϕ is a homomorphism, $\ker \phi = \{1\}$. ϕ is an epimorphism.
   c. ϕ is not a homomorphism.
   d. ϕ is a homomorphism, $\ker \phi = \{1, -1\}$. ϕ is not an epimorphism because there is no x such that $\phi(x) = -1$.
   e. ϕ is a homomorphism, $\ker \phi = \{x \mid x \in \mathbf{R}, x > 0\}$. ϕ is not an epimorphism since $\phi(G) = \{1, -1\}$.
   f. ϕ is not a homomorphism.
   g. ϕ is a homomorphism, $\ker \phi = \{1\}$. ϕ is an epimorphism.
   h. ϕ is not a homomorphism.

2. $\ker \phi = \langle [4] \rangle = \{[0], [4], [8]\}$. ϕ is not an epimorphism.

3. $\ker \phi = \{[0]_{12}, [6]_{12}\}$. ϕ is an epimorphism.

4. $\ker \phi = \{[0]_8, [6]_8\}$. ϕ is an epimorphism.

5. ϕ is an epimorphism.

6. An example is provided by G, G' and ϕ in Exercise 5 of this section.

10. Let G be the additive group $\mathbf{Z}$ and G' the additive group $\mathbf{Z}_n$.

14. The kernel of ϕ is the set of all multiples of the order of the element a in G.

Exercises 4.1

1. a. $(1,4)(2,5)$ b. $(2,3)(4,5)$ c. $(1,4,5,2)$ d. $(1,3,2,5)$
   e. $(1,3,5)(2,4,6)$ f. $(1,5,2)(4,7)$ g. $(1,4)(2,3,5)$
   h. $(1,2,4)(3,5)$

2. a. $(1,4,8,7,2,3)(5,9,6)$ b. $(1,2,9,3,4,5,6,7,8)$
   c. $(1,4,8,7)(2,6,5,3)$ d. $(1,5,4)(2,3)$ e. $(1,2)(3,4,5)$
   f. $(1,4)(2,3,8,7,6,5)$ g. $(1,7,6,4,3,5,2)$
   h. $(1,4,2,7)(3,5)(6,9)$

3. a. Even b. Even c. Odd d. Odd e. Even f. Odd
   g. Odd h. Odd

4. a. Odd b. Even c. Even d. Odd e. Odd f. Even
   g. Even h. Odd

5. a. Two b. Two c. Four d. Four e. Three f. Six
g. Six h. Six

6. a. Six b. Nine c. Four d. Six e. Six f. Six
g. Seven h. Four

7. a. $(1,4)(2,5)$ b. $(2,3)(4,5)$ c. $(1,2)(1,5)(1,4)$
d. $(1,5)(1,2)(1,3)$ e. $(1,5)(1,3)(2,6)(2,4)$ f. $(1,2)(1,5)(4,7)$
g. $(1,4)(2,5)(2,3)$ h. $(1,4)(1,2)(3,5)$

8. a. $(1,3)(1,2)(1,7)(1,8)(1,4)(5,6)(5,9)$
b. $(1,8)(1,7)(1,6)(1,5)(1,4)(1,3)(1,9)(1,2)$
c. $(1,7)(1,8)(1,4)(2,3)(2,5)(2,6)$ d. $(1,4)(1,5)(2,3)$
e. $(1,2)(3,5)(3,4)$ f. $(1,4)(2,5)(2,6)(2,7)(2,8)(2,3)$
g. $(1,2)(1,5)(1,3)(1,4)(1,6)(1,7)$ h. $(1,7)(1,2)(1,4)(3,5)(6,9)$

9. a. $f^2 = (1,2)(4,5)\,, f^3 = f^{-1}(1,4,2,5)$
b. $f^2 = (2,4,5,7,3)\,, f^3 = (2,3,7,5,4)\,, f^{-1} = (2,5,3,4,7)$
c. $f^2 = f^{-1} = (1,2,6)(3,5,4)\,, f^3 = (1)$
d. $f^2 = (3,7)(4,5)\,, f^3 = f^{-1} = (1,2)(3,4,7,5)$
e. $f^2 = (1,8,2)(3,7,6,4,5)\,, f^3 = (3,5,4,6,7)\,, f^{-1} = (1,8,2)(3,6,5,7,4)$
f. $f^2 = (1,7)(3,4)(2,9,6,5,8)\,, f^3 = (1,4,7,3)(2,8,5,6,9)\,,$
$f^{-1} = (1,4,7,3)(2,6,8,9,5)$

10. a. $(3,1,4,2) = (1,4,2,3)$ b. $(1,3,4,2)$ c. $(1,2,4,5)$
d. $(2,4)(3,1) = (2,4)(1,3)$ e. $(1,4,2)(5,3) = (1,4,2)(3,5)$
f. $(3,6,2,4)(5,1) = (2,4,3,6)(1,5)$

11. a. $(1,2)(4,9)(5,6)$ b. $(1,3,4)(5,6)(7,8)$ c. $(1,2)(3,4,5)$
d. $(1,2,3,4)(5,6)$ e. $(3,7,4,5)(6,8)$ f. $(2,3)(4,5)$

12. $g = f^3 = (1,4)(2,5)(3,6)\,, h = f^4 = (1,5,3)(2,6,4)$

13. $g = f^4 = (1,5,9)(2,6,10)(3,7,11)(4,8,12)\,,$
$h = f^9 = (1,10,7,4)(2,11,8,5)(3,12,9,6)$

14. $e = (1)\,, \rho = (1,2,3)\,, \rho^2 = (1,3,2)$

15.

$(1,2,3,4)$	$(1,2,3)$	$(1,2)$	$(1,2)(3,4)$
$(1,2,4,3)$	$(1,3,2)$	$(1,3)$	$(1,3)(2,4)$
$(1,3,2,4)$	$(1,2,4)$	$(1,4)$	$(1,4)(2,3)$
$(1,3,4,2)$	$(1,4,2)$	$(2,3)$	(1)
$(1,4,2,3)$	$(1,3,4)$	$(2,4)$	
$(1,4,3,2)$	$(1,4,3)$	$(3,4)$	
	$(2,3,4)$		
	$(2,4,3)$		

16. $\langle(1)\rangle = \{(1)\}$

$\langle(1,2,3)\rangle = \langle(1,3,2)\rangle = \{(1), (1,2,3), (1,3,2)\}$

$\langle(1,2,4)\rangle = \langle(1,4,2)\rangle = \{(1), (1,2,4), (1,4,2)\}$

$\langle(1,4,3)\rangle = \langle(1,3,4)\rangle = \{(1), (1,4,3), (1,3,4)\}$

$\langle(2,3,4)\rangle = \langle(2,4,3)\rangle = \{(1), (2,3,4), (2,4,3)\}$

$\langle(1,2)(3,4)\rangle = \{(1), (1,2)(3,4)\}$

$\langle(1,3)(2,4)\rangle = \{(1), (1,3)(2,4)\}$

$\langle(1,4)(2,3)\rangle = \{(1), (1,4)(2,3)\}$

17. $\langle(1,2)\rangle = \{(1), (1,12)\}$ has order 2.

$\langle(1,2,3)\rangle = \{(1), (1,2,3), (1,3,2)\}$ has order 3.

$\langle(1,2,3,4)\rangle = \{(1), (1,2,3,4), (1,3)(2,4), (1,4,3,2)\}$ has order 4.

18.

$\circ$	e	α	α^2	α^3	β	γ	$\triangle$	θ
e	e	α	α^2	α^3	β	γ	$\triangle$	θ
α	α	α^2	α^3	e	γ	$\triangle$	θ	β
α^2	α^2	α^3	e	α	$\triangle$	θ	β	γ
α^3	α^3	e	α	α^2	θ	β	γ	$\triangle$
β	β	θ	$\triangle$	γ	e	α^3	α^2	α
γ	γ	β	θ	$\triangle$	α	e	α^3	α^2
$\triangle$	$\triangle$	γ	β	θ	α^2	α	e	α^3
θ	θ	$\triangle$	γ	β	α^3	α^2	α	e

19. $\{e\}, \{e,\beta\}, \{e,\gamma\}, \{e,\triangle\}, \{e,\theta\}, \{e,\alpha^2\}, \{e,\alpha,\alpha^2,\alpha^3\}$

20. Let $\phi : G \to G'$ be defined by

$$\phi(e) = I_2, \ \phi(\alpha) = R, \ \phi(\alpha^2) = R^2, \ \phi(\alpha^3) = R^3,$$
$$\phi(\beta) = H, \ \phi(\gamma) = D, \ \phi(\triangle) = V, \ \phi(\theta) = T.$$

22. $C_e = C_{\alpha^2} = G$
$C_\alpha = C_{\alpha^3} = \{e, \alpha, \alpha^2, \alpha^3\}$
$C_\beta = C_\triangle = \{e, \alpha^2, \beta, \triangle\}$
$C_\gamma = C_\theta = \{e, \alpha^2, \gamma, \theta\}$

Exercises 4.2

1. With $f_g : G \to G$ defined by $f_g(x) = gx$ for each $g \in G$, we obtain the following permutations on the set of elements of G :

$$f_e = (e), \quad f_a = (e, a)(b, ab), \quad f_b = (e, b)(a, ab), \quad f_{ab} = (e, ab)(a, b).$$

The set $G' = \{f_e, f_a, f_b, f_{ab}\}$ is a group of permutations, and the mapping $\phi : G \to G'$ defined by

$$\phi : \begin{cases} \phi(e) = f_e \\ \phi(a) = f_a \\ \phi(b) = f_b \\ \phi(ab) = f_{ab} \end{cases}$$

is an isomorphism from G to G'.

2. For notational convenience we let a represent $[a]$ in this solution. With $f_g : G \to G$ defined by $f_g(x) = gx$ for each $g \in G$, we obtain the following permutations on the set of elements of G :

$$f_1 = (1), \quad f_2 = (1, 2, 4, 3), \quad f_3 = (1, 3, 4, 2), \quad f_4 = (1, 4)(2, 3).$$

The set $G' = \{f_1, f_2, f_3, f_4\}$ is a group of permutations, and the mapping $\phi : G \to G'$ defined by

$$\phi : \begin{cases} \phi(1) = f_1 \\ \phi(2) = f_2 \\ \phi(3) = f_3 \\ \phi(4) = f_4 \end{cases}$$

is an isomorphism from G to G'.

3. For notational convenience we let a represent $[a]$ in this solution. Let $f_a : G \to G$ be defined by $f_a(x) = ax$ for each $x \in G$. Then we have the following permutations:

$$f_2 = (2, 4, 8, 6), \quad f_4 = (2, 8)(4, 6), \quad f_6 = (6), \quad f_8 = (2, 6, 8, 4).$$

The set $G' = \{f_2, f_4, f_6, f_8\}$ is a group of permutations, and the mapping $\phi : G \to G'$ defined by

$$\phi : \begin{cases} \phi(2) = f_2 \\ \phi(4) = f_4 \\ \phi(6) = f_6 \\ \phi(8) = f_8 \end{cases}$$

is an isomorphism from G to G'.

4. With $f_a : G \to G$ defined by $f_a(x) = ax$, for each $x \in G$, we obtain the following permutations on the set of elements of G.

$$\begin{array}{ll} f_{I_3} = (I_3) & f_{P_3} = (I_3, P_3, P_5)(P_1, P_4, P_2) \\ f_{P_1} = (I_3, P_1)(P_2, P_3)(P_4, P_5) & f_{P_4} = (I_3, P_4)(P_1, P_3)(P_2, P_5) \\ f_{P_2} = (I_3, P_2)(P_1, P_5)(P_3, P_4) & f_{P_5} = (I_3, P_5, P_3)(P_1, P_2, P_4) \end{array}$$

The set $G' = \{f_{I_3}, f_{P_1}, f_{P_2}, f_{P_3}, f_{P_4}, f_{P_5}\}$ is a group of permutations, and the mapping $\phi : G \to G'$ defined by

$$\begin{array}{ll} \phi(I_3) = f_{I_3} & \phi(P_3) = f_{P_3} \\ \phi(P_1) = f_{P_1} & \phi(P_4) = f_{P_4} \\ \phi(P_2) = f_{P_2} & \phi(P_5) = f_{P_5} \end{array}$$

is an isomorphism from G to G'.

5. c. The mapping ϕ is not an isomorphism when G is not abelian.

6. c. The mapping ϕ is an isomorphism.

7. c. The mapping ϕ is not an isomorphism when G is not abelian.

Exercises 4.3

1. $\{I, V\}$, where I is the identity mapping and V is the reflection about the vertical axis of symmetry.

2. $\{I, V\}$, where I is the identity mapping and V is the reflection about the vertical axis of symmetry.

3. $\{I, R\}$, where I is the identity mapping and R is the counterclockwise rotation through 180° about the center of symmetry.

4. $\{I, V, H, R\}$, where I is the identity mapping, V is the reflection about the vertical axis of symmetry, H is the reflection about the horizontal axis of symmetry, and R is the counterclockwise rotation through 180° about the center of symmetry.

5. Rotational symmetry only

6. Neither rotational symmetry nor reflective symmetry

7. Reflective symmetry only **8**. Rotational symmetry only

9. Both rotational symmetry and reflective symmetry

10. Both rotational symmetry and reflective symmetry

11. $\{R, R^2, R^3 = I\}$, where I is the identity mapping and R is the counterclockwise rotation through 120° about the center of the triangle determined by the arrow tips.

12. $\{R, R^2, R^3, R^4, R^5 = I\}$, where I is the identity mapping and R is the counterclockwise rotation through 72° about the center of the pentagon determined by the tips of the star.

13. Let the vertices of the ellipses be numbered as in the following figure.

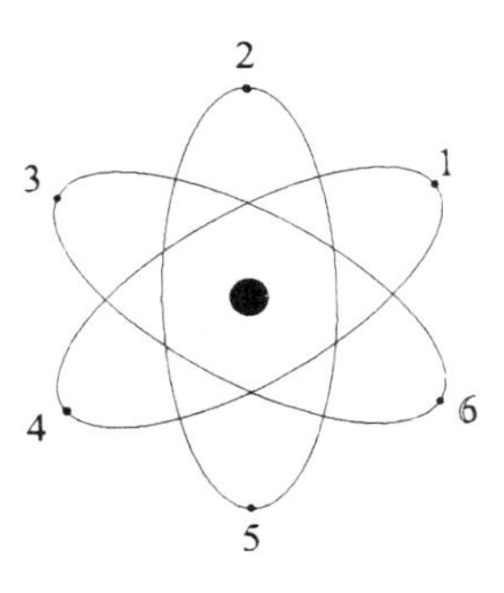

Then any symmetry of the figure can be identified with the corresponding permutation on $\{1, 2, 3, 4, 5, 6\}$, and the group G of symmetries of the figure can be described with the notation

$$G = \{R, R^2, R^3, R^4, R^5, R^6 = I, L, LR, LR^2, LR^3, LR^4, LR^5\},$$

where

$$\begin{aligned} I &= (1) & L &= (2,6)(3,5) \\ R &= (1,2,3,4,5,6) & LR &= (1,6)(2,5)(3,4) \\ R^2 &= (1,3,5)(2,4,6) & LR^2 &= (1,5)(2,4) \\ R^3 &= (1,4)(2,5)(3,6) & LR^3 &= (1,4)(2,3)(5,6) \\ R^4 &= (1,5,3)(2,6,4) & LR^4 &= (1,3)(4,6) \\ R^5 &= (1,6,5,4,3,2) & LR^5 &= (1,2)(3,6)(4,5)\,. \end{aligned}$$

This is the same permutation group as the one in the answer to Exercise 24 of this exercise set.

14. Let the axes of symmetry be labeled as in the following figure.

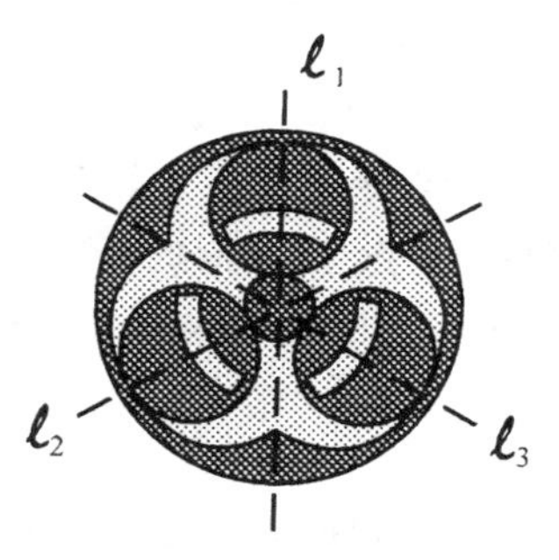

Then the group G of symmetries of the figure can be described as

$$G = \{I, R, R^2, L, LR, LR^2\}\,,$$

where

I is the identity mapping,

R is the rotation through 120° counterclockwise about the center,

R^2 is the rotation through 240° counterclockwise about the center,

L is the reflection about the vertical axis ℓ_1,

LR is the reflection about the axis ℓ_2, and

LR^2 is the relection about the axis ℓ_3.

15. Let the axes of symmetry be labeled as in the following figure.

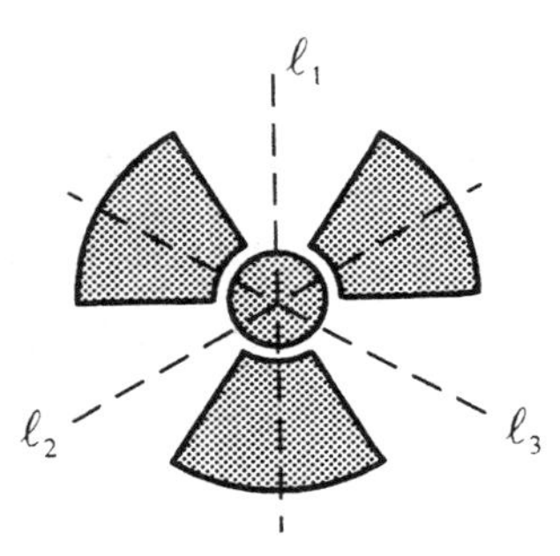

Then the group G of symmetries of the figure can be described as

$$G = \{I, R, R^2, L, LR, LR^2\},$$

where

I is the identity mapping,

R is the rotation through 120° counterclockwise about the center,

R^2 is the rotation through 240° counterclockwise about the center,

L is the reflection about the vertical axis ℓ_1,

LR is the reflection about the axis ℓ_2, and

LR^2 is the relection about the axis ℓ_3.

16. $\{R, R^2 = I, V, H\}$, where R is the counterclockwise rotation through 180° about the center of symmetry, V is the reflection about the vertical axis of symmetry, and H is the reflection about the horizontal axis of symmetry.

17. Let I denote the identity mapping, and let t denote a translation of the set of E's one unit to the right. Then t^{-1} is a translation of the set of E's one unit to the left, and the collection

$$\{\ldots, t^{-2}, t^{-1}, t^0 = I, t, t^2, \ldots\}$$

are elements of the (infinite) group of symmetries of the figure. Let r denote the reflection of the figure about the horizontal axis of symmetry through the E's. Then $r^2 = I = r^0, rt = tr$, and the group of symmetries consists of all products of the form $r^i t^j$, where i is either 0 or 1 and j is an integer.

18. Let I denote the identity mapping, and let t denote a translation of the set of $\triangleright$'s one unit to the right. Then t^{-1} is a translation of the set of $\triangleright$'s one unit to the left, and the collection

$$\{\ldots, t^{-2}, t^{-1}, t^0 = I, t, t^2, \ldots\}$$

are elements of the (infinite) group of symmetries of the figure. Let r denote the reflection of the figure about the horizontal axis of symmetry through the $\triangleright$'s. Then $r^2 = I = r^0$, $rt = tr$, and the group of symmetries consists of all products of the form $r^i t^j$, where i is either 0 or 1 and j is an integer.

19. Let I denote the identity mapping, and let t denote a translation of the set of **T**'s one unit to the right. Then t^{-1} is a translation of the set of **T**'s one unit to the left. There is a vertical axis of symmetry through each copy of the letter **T**, and a corresponding reflection of the figure about that vertical axis. Each of these reflections is its own inverse. The group of symmetries consists of this infinite collection of reflections (one for each copy of the letter **T**) together with the identity I and all the integral powers of the translation t.

20. Let I denote the identity mapping and let t denote a translation of the set of stars one unit to the right. Then t^{-1} is a translation of the set of stars one unit to the left. There is a vertical axis of symmetry through each star, and a corresponding reflection of the figure about that vertical axis. Each of these reflections is its own inverse. The group of symmetries consists of this infinite collection of reflections (one for each star) together with the identity I and all the integral powers of the translation t.

22. Using the same notational convention as in Example 11 of Section 4.1, the elements of G are as follows:

$$e = (1)\,, \quad \alpha = (1,3)\,(2,4)\,, \quad \beta = (1,4)\,(2,3)\,, \quad \triangle = (1,2)\,(3,4)\,.$$

With this notation, we obtain the following multiplication table for G.

$\circ$	e	α	β	$\triangle$
e	e	α	β	$\triangle$
α	α	e	$\triangle$	β
β	β	$\triangle$	e	α
$\triangle$	$\triangle$	β	α	e

23. Using the same notational convention as in Example 11 of Section 4.1, the elements of G are as follows:

$$\begin{aligned} e &= (1) & \beta &= (2,5)\,(3,4) \\ \alpha &= (1,2,3,4,5) & \gamma &= \alpha\beta = \beta\alpha^4 = (1,2)\,(3,5) \\ \alpha^2 &= (1,3,5,2,4) & \triangle &= \alpha^2\beta = \beta\alpha^3 = (1,3)\,(4,5) \\ \alpha^3 &= (1,4,2,5,3) & \theta &= \alpha^3\beta = \beta\alpha^2 = (1,4)\,(2,3) \\ \alpha^4 &= (1,5,4,3,2) & \sigma &= \alpha^4\beta = \beta\alpha = (1,5)\,(2,4)\,. \end{aligned}$$

With this notation, we obtain the following multiplication table for G.

$\circ$	e	α	α^2	α^3	α^4	β	γ	$\triangle$	θ	σ
e	e	α	α^2	α^3	α^4	β	γ	$\triangle$	θ	σ
α	α	α^2	α^3	α^4	e	γ	$\triangle$	θ	σ	β
α^2	α^2	α^3	α^4	e	α	$\triangle$	θ	σ	β	γ
α^3	α^3	α^4	e	α	α^2	θ	σ	β	γ	$\triangle$
α^4	α^4	e	α	α^2	α^3	σ	β	γ	$\triangle$	θ
β	β	σ	θ	$\triangle$	γ	e	α^4	α^3	α^2	α
γ	γ	β	σ	θ	$\triangle$	α	e	α^4	α^3	α^2
$\triangle$	$\triangle$	γ	β	σ	θ	α^2	α	e	α^4	α^3
θ	θ	$\triangle$	γ	β	σ	α^3	α^2	α	e	α^4
σ	σ	θ	$\triangle$	γ	β	α^4	α^3	α^2	α	e

24. Using the same notational convention as in Example 11 of Section 4.1, the elements of G are as follows:

(1) $\qquad (1,4)(2,3)(5,6)$

$(1,2,3,4,5,6)$ $\qquad (1,6)(2,5)(3,4)$

$(1,3,5)(2,4,6)$ $\qquad (1,2)(3,6)(4,5)$

$(1,4)(2,5)(3,6)$ $\qquad (2,6)(3,5)$

$(1,5,3)(2,6,4)$ $\qquad (1,3)(4,6)$

$(1,6,5,4,3,2)$ $\qquad (1,5)(2,4)$

25. 48 **26.** 24

27. Using the same notational convention as in Example 11 of Section 4.1, the elements of $G = \{e, \alpha, \beta, \triangle\}$ are given by $e = (1), \alpha = (1,3)(2,4), \beta = (1,4)(2,3), \triangle = (1,2)(3,4)$. Let $\phi : G \to H$ be defined by

$$\phi(e) = \begin{bmatrix} 1 & 0 \\ 0 & 1 \end{bmatrix}, \quad \phi(\alpha) = \begin{bmatrix} 1 & 0 \\ 0 & -1 \end{bmatrix},$$

$$\phi(\beta) = \begin{bmatrix} -1 & 0 \\ 0 & 1 \end{bmatrix}, \quad \phi(\triangle) = \begin{bmatrix} -1 & 0 \\ 0 & -1 \end{bmatrix}.$$

Exercises 4.4

1. a. $eH = \beta H = \{e, \beta\}$; $\alpha H = \gamma H = \{\alpha, \gamma\}$; $\alpha^2 H = \Delta H = \{\alpha^2, \Delta\}$; $\alpha^3 H = \theta H = \{\alpha^3, \theta\}$

   b. $He = H\beta = \{e, \beta\}$; $H\alpha = H\theta = \{\alpha, \theta\}$; $H\alpha^2 = H\Delta = \{\alpha^2, \Delta\}$; $H\alpha^3 = H\gamma = \{\alpha^3, \gamma\}$

2. a. $eH = \Delta H = \{e, \Delta\}$; $\alpha H = \theta H = \{\alpha, \theta\}$; $\alpha^2 H = \beta H = \{\alpha^2, \beta\}$; $\alpha^3 H = \gamma H = \{\alpha^3, \gamma\}$

   b. $He = H\Delta = \{e, \Delta\}$; $H\alpha = H\gamma = \{\alpha, \gamma\}$; $H\alpha^2 = H\beta = \{\alpha^2, \beta\}$; $H\alpha^3 = H\theta = \{\alpha^3, \theta\}$

3. a. $I_3 H = P_4 H = \{I_3, P_4\}$; $P_1 H = P_3^2 H = \{P_1, P_3^2\}$; $P_2 H = P_3 H = \{P_2, P_3\}$

   b. $HI_3 = HP_4 = \{I_3, P_4\}$; $HP_1 = HP_3 = \{P_1, P_3\}$; $HP_2 = HP_3^2 = \{P_2, P_3^2\}$

12. a. $\{e, \alpha^2\}$ b. $\{e, \Delta\}$

14. Order 1: $\{e\}$
    Order 2: $\{e, \alpha^2\}, \{e, \beta\}, \{e, \gamma\}, \{e, \Delta\}, \{e, \theta\}$
    Order 4: $\{e, \alpha, \alpha^2, \alpha^3\}, \{e, \beta, \Delta, \alpha^2\}, \{e, \gamma, \theta, \alpha^2\}$
    Order 8: $\{e, \alpha, \alpha^2, \alpha^3, \beta, \gamma, \Delta, \theta\}$

15. Order 1: $\{(1)\}$

    Order 2: $\{(1), (1,2)(3,4)\}$, $\{(1), (1,3)(2,4)\}$, $\{(1), (1,4)(2,3)\}$

    Order 3: $\{(1), (1,2,3), (1,3,2)\}$, $\{(1), (1,2,4), (1,4,2)\}$, $\{(1), (1,4,3), (1,3,4)\}$, $\{(1), (2,3,4), (2,4,3)\}$

    Order 4: $\{(1), (1,2)(3,4), (1,3)(2,4), (1,4)(2,3)\}$

    Order 12: A_4, as given in Example 8 of Section 4.1

16. Order 1: $\{1\}$
    Order 2: $\{-1, 1\}$
    Order 4: $\{i, -1, -i, 1\}, \{j, -1, -j, 1\}, \{k, -1, -k, 1\}$
    Order 8: $\{1, -1, i, -i, j, -j, k, -k\}$

17. The normal subgroups of the octic group G are $H_1 = \{e\}, H_2 = \{e, \alpha^2\}, H_3 = \{e, \alpha, \alpha^2, \alpha^3\}$, $H_4 = \{e, \beta, \Delta, \alpha^2\}, H_5 = \{e, \gamma, \theta, \alpha^2\}$, and $H_6 = G$.

18. Order 1: $\{(1)\}$
    Order 4: $\{(1), (1,2)(3,4), (1,3)(2,4), (1,4)(2,3)\}$
    Order 12: A_4, as given in Example 8 of Section 4.1.

19. The normal subgroups of the quaternion group G are $H_1 = \{1\}, H_2 = \{-1, 1\}, H_3 = \{i, -1, -i, 1\}$, $H_4 = \{j, -1, -j, 1\}, H_5 = \{k, -1, -k, 1\}$, and $H_6 = G$.

20. a. H is a normal subgroup of G, where $H = \{(1), (1,2)(3,4)\}$ and $G = \{(1), (1,2)(3,4), (1,3)(2,4), (1,4)(2,3)\}$.

b. This G is a normal subgroup of A_4.

c. This H is not a normal subgroup of A_4.

21. $H = \{e, \triangle\}, K = \{e, \beta, \triangle, \alpha^2\}$

22. The group S_3 of Example 6 in Section 4.1 and any cyclic group of order 6.

33. $\{e, \alpha^2\}$ **34**. $\{(1)\}$

36. For $H = \{(1), (1,3)(2,4)\}, \mathcal{N}(H)$ is the octic group G since H is normal in G.

37. $\{(1), (1,4)(2,3), (1,2)(3,4), (1,3)(2,4)\}$

39. The equivalence classes are the same as the left cosets of H in G. That is, each equivalence class is a left coset of H in G, and each left coset of H in G is an equivalence class.

49. a. $S_3 = \{(1), (1,2), (1,3), (2,3), (1,2,3), (1,3,2)\}$

b. $\{(1), (1,3), (1,2,3,4), (1,3)(2,4), (1,4,3,2), (1,2)(3,4), (2,4), (1,4)(2,3)\}$

c. A_4, as given in Example 8 of Section 4.1.

d. S_4

Exercises 4.5

1. $G/H = \{H, \alpha H, \beta H, \gamma H\}$ where $H = \{e, \alpha^2\}, \alpha H = \{\alpha, \alpha^3\}, \beta H = \{\beta, \triangle\}, \gamma H = \{\gamma, \theta\}$.

$\cdot$	H	αH	βH	γH
H	H	αH	βH	γH
αH	αH	H	γH	βH
βH	βH	γH	H	αH
γH	γH	βH	αH	H

2. $G/H = \{H, (1,2,3)H, (1,3,2)H\}$, where

$$(1,2,3)H = \{(1,2,3), (1,3,4), (2,4,3), (1,4,2)\},$$
$$(1,3,2)H = \{(1,3,2), (2,3,4), (1,2,4), (1,4,3)\}.$$

$\cdot$	H	$(1,2,3)\,H$	$(1,3,2)\,H$
H	H	$(1,2,3)\,H$	$(1,3,2)\,H$
$(1,2,3)\,H$	$(1,2,3)\,H$	$(1,3,2)\,H$	H
$(1,3,2)\,H$	$(1,3,2)\,H$	H	$(1,2,3)\,H$

3. $G/H = \{H, iH, jH, kH\}$ where $H = \{1, -1\}, iH = \{i, -i\}, jH = \{j, -j\}, kH = \{k, -k\}$.

$\cdot$	H	iH	jH	kH
H	H	iH	jH	kH
iH	iH	H	kH	jH
jH	jH	kH	H	iH
kH	kH	jH	iH	H

4. $H = \{[1], [9]\}, G/H = \{H, [3]\,H, [11]\,H, [13]\,H\}$, where

$$[3]\,H = \{[3], [7]\},$$
$$[11]\,H = \{[11], [19]\},$$
$$[13]\,H = \{[13], [17]\}.$$

$\cdot$	H	$[3]\,H$	$[11]\,H$	$[13]\,H$
H	H	$[3]\,H$	$[11]\,H$	$[13]\,H$
$[3]\,H$	$[3]\,H$	H	$[13]\,H$	$[11]\,H$
$[11]\,H$	$[11]\,H$	$[13]\,H$	H	$[3]\,H$
$[13]\,H$	$[13]\,H$	$[11]\,H$	$[3]\,H$	H

5. The normal subgroups of the octic group G are $H_1 = \{e\}, H_2 = \{e, \alpha^2\}, H_3 = \{e, \alpha, \alpha^2, \alpha^3\}$, $H_4 = \{e, \beta, \Delta, \alpha^2\}, H_5 = \{e, \gamma, \theta, \alpha^2\}$, and $H_6 = G$. We consider the possible quotient groups.

(1) G/H_1 is isomorphic to G.

(2) $G/H_2 = \{H_2, \alpha H_2, \beta H_2, \gamma H_2\}$ is isomorphic to the Klein four group. (See Exercise 1 of Section 4.2.)

(3) Each of $G/H_3, G/H_4$, and G/H_5 is a cyclic group of order 2.

(4) $G/G = \{G\}$ is a group of order 1.

Thus, the homomorphic images of the octic group G are G itself, a Klein four group, a cyclic group of order 2, and a group with only the identity element.

6. The normal subgroups of A_4 are H_1, H_2, and H_3, where

$$H_1 = \{(1)\},$$
$$H_2 = \{(1), (1,2)(3,4), (1,3)(2,4), (1,4)(2,3)\},$$
$$H_3 = A_4.$$

We consider the possible quotient groups.

(1) G/H_1 is isomorphic to G.

(2) G/H_2 is a cyclic group of order 3. (See Exercise 2 of this section.)

(3) G/H_3 is a group of order 1.

Thus the homomorphic images of G are the group G itself, a cyclic group of order 3, and a group with only the identity element.

7. The normal subgroups of the quaternion group G are $H_1 = \{1\}, H_2 = \{-1, 1\}, H_3 = \{i, -1, -i, 1\}$, $H_4 = \{j, -1, -j, 1\}, H_5 = \{k, -1, -k, 1\}$, and $H_6 = G$. We consider the quotient groups.

(1) G/H_1 is isomorphic to G.

(2) $G/H_2 = \{H_2, iH_2, jH_2, kH_2\}$ is isomorphic to the Klein four group. (See Exercise 1 of Section 4.2.)

(3) Each of $G/H_3, G/H_4$, and G/H_5 is a cyclic group of order 2.

(4) $G/G = \{G\}$ is a group of order 1.

Thus, the homomorphic images of the quaternion group G are G itself, a Klein four group, a cyclic group of order 2, and a group with only the identity element.

8. **a**. We have $G = \{[1], [3], [7], [9], [11], [13], [17], [19]\}$. The normal subgroups of G are

$$H_1 = \{[1]\}, \qquad H_5 = \{[1], [3], [9], [7]\},$$
$$H_2 = \{[1], [9]\}, \qquad H_6 = \{[1], [13], [9], [17]\},$$
$$H_3 = \{[1], [11]\}, \qquad H_7 = G.$$
$$H_4 = \{[1], [19]\},$$

The homomorphic images are the possible quotient groups.

(1) G/H_1 is isomorphic to G.

(2) Each of G/H_3 and G/H_4 is a cyclic group of order 4.

(3) G/H_2 is isomorphic to the Klein four group. (See Exercise 4 of this section.)

(4) Each of G/H_5 and G/H_6 is a cyclic group of order 2.

(5) $G/G = \{G\}$ is a group of order 1.

Thus the homomorphic images of G are the group G itself, a cyclic group of order 4, a Klein four group, a cyclic group of order 2, and a group with only the identity element.

b. We have $G = \{[1], [3], [5], [7]\}$. The normal subgroups of G are

$$H_1 = \{[1]\}, \qquad H_4 = \{[1], [7]\},$$
$$H_2 = \{[1], [3]\}, \qquad H_5 = G.$$
$$H_3 = \{[1], [5]\},$$

The homomorphic images are the possible quotient groups.

(1) G/H_1 is isomorphic to G.

(2) Each of $G/H_2, G/H_3$ and G/H_4 is a cyclic group of order 2.

(3) $G/G = \{G\}$ is a group of order 1.

Thus the homomorphic images of G are the group G itself, a cyclic group of order 2 and a group with only the identity element.

c. The group G is given by $G = \{[1], [5], [7], [11], [13], [17], [19], [23]\}$. The trivial subgroups of G are $H_1 = \{[1]\}$ and G itself. Every element of G except $[1]$ has order 2, so there are seven cyclic subgroups $H_2, H_3, \ldots, H_8$ of order 2 in G. There are seven subgroups of G that are isomorphic to the Klein four group:

$$H_9 = \{[1], [5], [7], [11]\}, \qquad H_{13} = \{[1], [7], [17], [23]\},$$
$$H_{10} = \{[1], [5], [13], [17]\}, \qquad H_{14} = \{[1], [11], [13], [23]\},$$
$$H_{11} = \{[1], [5], [19], [23]\}, \qquad H_{15} = \{[1], [11], [17], [19]\}.$$
$$H_{12} = \{[1], [7], [13], [19]\},$$

The homomorphic images of G are the possible quotient groups.

(1) G/H_1 is isomorphic to G.

(2) Each of $G/H_2, G/H_3, \ldots, G/H_8$ is isomorphic to the Klein four group.

(3) Each of $G/H_9, G/H_{10}, \ldots, G/H_{15}$ is a cyclic group of order 2.

(4) $G/G = \{G\}$ is a group of order 1.

Thus the homomorphic images of G are the group G itself, a Klein four group, a cyclic group of order 2, and a group with only the identity element.

d. The elements of G are given by $G = \{[1], [7], [11], [13], [17], [19], [23], [29]\}$. The normal subgroups of G are

$$H_1 = \{[1]\}, \qquad H_5 = \{[1], [7], [19], [13]\},$$
$$H_2 = \{[1], [11]\}, \qquad H_6 = \{[1], [17], [19], [23]\},$$
$$H_3 = \{[1], [29]\}, \qquad H_7 = G.$$
$$H_4 = \{[1], [19]\},$$

The homomorphic images are the possible quotient groups.

(1) G/H_1 is isomorphic to G.
(2) Each of G/H_2 and G/H_3 is a cyclic group of order 4.
(3) G/H_4 is isomorphic to the Klein four group.
(4) Each of G/H_5 and G/H_6 is a cyclic group of order 2.
(5) $G/G = \{G\}$ is a group of order 1.

Thus the homomorphic images of G are the group G itself, a cyclic group of order 4, a Klein four group, a cyclic group of order 2, and a group with only the identity element.

9. **a**. The left cosets of $H = \{(1), (1,2)\}$ in $G = S_3$ are given by

$$(1)\,H = (1,2)\,H = \{(1), (1,2)\},$$
$$(1,3)\,H = (1,2,3)\,H = \{(1,3), (1,2,3)\},$$
$$(2,3)\,H = (1,3,2)\,H = \{(2,3), (1,3,2)\}.$$

The rule $aHbH = abH$ leads to

$$(1,3)\,H\,(2,3)\,H = (1,3)\,(2,3)\,H = (1,3,2)\,H$$

and also to

$$(1,2,3)\,H\,(1,3,2)\,H = (1,2,3)\,(1,3,2)\,H = (1)\,H.$$

We have $(1,3)\,H = (1,2,3)\,H$ and $(2,3)\,H = (1,3,2)\,H$, but

$$(1,3)\,H\,(2,3)\,H \neq (1,2,3)\,H\,(1,3,2)\,H.$$

Thus the rule $aHbH = abH$ does not define a binary operation on the left cosets of H in G. (That is, the result is not well-defined.)

b. The left cosets of $H = \{(1), (1,3)\}$ in $G = S_3$ are given by

$$(1)\,H = (1,3)\,H = \{(1), (1,3)\},$$
$$(1,2)\,H = (1,3,2)\,H = \{(1,2), (1,3,2)\},$$
$$(2,3)\,H = (1,2,3)\,H = \{(2,3), (1,2,3)\}.$$

The rule $aHbH = abH$ leads to

$$(1,2)\,H\,(2,3)\,H = (1,2)\,(2,3)\,H = (1,2,3)\,H$$

and also to

$$(1,3,2)\,H\,(1,2,3)\,H = (1,3,2)\,(1,2,3)\,H = (1)\,H.$$

We have $(1,2)\,H = (1,3,2)\,H$ and $(2,3)\,H = (1,2,3)\,H$, but

$$(1,2)\,H\,(2,3)\,H \neq (1,3,2)\,H\,(1,2,3)\,H.$$

Thus the rule $aHbH = abH$ does not define a binary operation on the left cosets of H in G. (That is, the result is not well-defined.)

c. The left cosets of $H = \{(1), (2,3)\}$ in $G = S_3$ are given by

$$\begin{aligned}(1)\,H = (2,3)\,H &= \{(1), (2,3)\},\\ (1,2)\,H = (1,2,3)\,H &= \{(1,2), (1,2,3)\},\\ (1,3)\,H = (1,3,2)\,H &= \{(1,3), (1,3,2)\}.\end{aligned}$$

The rule $aHbH = abH$ leads to

$$(1,2)\,H\,(1,3)\,H = (1,2)\,(1,3)\,H = (1,3,2)\,H$$

and also to

$$(1,2,3)\,H\,(1,3,2)\,H = (1,2,3)\,(1,3,2)\,H = (1)\,H.$$

We have $(1,2)\,H = (1,2,3)\,H$ and $(1,3)\,H = (1,3,2)\,H$, but

$$(1,2)\,H\,(1,3)\,H \neq (1,2,3)\,H\,(1,3,2)\,H.$$

Thus the rule $aHbH = abH$ does not define a binary operation on the left cosets of H in G. (That is, the result is not well-defined.)

10. The distinct cosets of H in S_4 are as follows.

$$\begin{aligned}H &= \{(1), (1,2)(3,4), (1,3)(2,4), (1,4)(2,3)\}\\ (1,2)\,H &= \{(1,2), (3,4), (1,3,2,4), (1,4,2,3)\}\\ (1,3)\,H &= \{(1,3), (2,4), (1,2,3,4), (1,4,3,2)\}\\ (1,4)\,H &= \{(1,4), (2,3), (1,2,4,3), (1,3,4,2)\}\\ (1,2,3)\,H &= \{(1,2,3), (1,3,4), (2,4,3), (1,4,2)\}\\ (1,3,2)\,H &= \{(1,3,2), (1,4,3), (2,3,4), (1,2,4)\}\end{aligned}$$

$\cdot$	H	$(1,2)\,H$	$(1,3)\,H$	$(1,4)\,H$	$(1,2,3)\,H$	$(1,3,2)\,H$
H	H	$(1,2)\,H$	$(1,3)\,H$	$(1,4)\,H$	$(1,2,3)\,H$	$(1,3,2)\,H$
$(1,2)\,H$	$(1,2)\,H$	H	$(1,3,2)\,H$	$(1,2,3)\,H$	$(1,4)\,H$	$(1,3)\,H$
$(1,3)\,H$	$(1,3)\,H$	$(1,2,3)\,H$	H	$(1,3,2)\,H$	$(1,2)\,H$	$(1,4)\,H$
$(1,4)\,H$	$(1,4)\,H$	$(1,3,2)\,H$	$(1,2,3)\,H$	H	$(1,3)\,H$	$(1,2)\,H$
$(1,2,3)\,H$	$(1,2,3)\,H$	$(1,3)\,H$	$(1,4)\,H$	$(1,2)\,H$	$(1,3,2)\,H$	H
$(1,3,2)\,H$	$(1,3,2)\,H$	$(1,4)\,H$	$(1,2)\,H$	$(1,3)\,H$	H	$(1,2,3)\,H$

11. The mapping ϕ is a homomorphism, and $\ker\phi = A_n$.

12. a. $K = \{I, R, R^2, R^3\}$ b. $K = \{I, R, R^2, R^3\}, HK = \{H, V, D, T\}$
c. $\theta(K) = 1, \quad \theta(HK) = -1$

13. a. $K = \{I_2, M_2, M_3\}$ b. $K = \{I_2, M_2, M_3\}, \; M_1K = \{M_1, M_4, M_5\}$
c. $\theta(K) = 1, \quad \theta(M_1K) = -1$

14. a. $K = \{1, -1\}$
b. $K = \{1, -1\}, \; iK = \{i, -i\}, \; jK = \{j, -j\}, \; kK = \{k, -k\}$
c. $\theta(K) = e, \quad \theta(iK) = a, \quad \theta(jK) = b, \quad \theta(kK) = ab$

15. a. $K = \{[1], [11]\}$
b. $K = \{[1], [11]\}, \; [3]K = \{[3], [13]\}, \; [7]K = \{[7], [17]\}, \; [9]K = \{[9], [19]\}$
c. $\theta(K) = e, \quad \theta([3]K) = a, \quad \theta([7]K) = a^3, \quad \theta([9]K) = a^2$

20. a. Let $G = \{a, a^2, a^3, a^4, a^5, a^6, a^7, a^8 = e\}$ be a cyclic group of order 8. The subgroup $H = \{a^2, a^4, a^6, a^8 = e\}$ of G is a cyclic group of order 4, and the mapping $\phi : G \to H$ defined by $\phi(x) = x^2$ is a homomorphism since

$$\begin{aligned} \phi(xy) &= (xy)^2 \\ &= x^2y^2 \quad \text{since } G \text{ is abelian} \\ &= \phi(x)\,\phi(y). \end{aligned}$$

The mapping ϕ is an epimorphism since

$$\begin{aligned} \phi(G) &= \{\phi(a), \phi(a^2), \phi(a^3), \phi(a^4), \phi(a^5), \phi(a^6), \phi(a^7), \phi(e)\} \\ &= \{a^2, a^4, a^6, a^8 = e, a^{10} = a^2, a^{12} = a^4, a^{14} = a^6, e\} \\ &= \{a^2, a^4, a^6, a^8 = e\} \\ &= H. \end{aligned}$$

Thus, G has H as a homomorphic image.

b. Let $G = \{a, a^2, a^3, a^4, a^5, a^6 = e\}$ be a cyclic group of order 6. The subgroup $H = \{a^3, a^6 = e\}$ of G is a cyclic group of order 2, and the mapping $\phi : G \to H$ defined by $\phi(x) = x^3$ is a homomorphism since

$$\begin{aligned} \phi(xy) &= (xy)^3 \\ &= x^3y^3 \quad \text{since } G \text{ is abelian} \\ &= \phi(x)\phi(y). \end{aligned}$$

The mapping ϕ is an epimorphism since

$$\begin{aligned} \phi(G) &= \{\phi(a), \phi(a^2), \phi(a^3), \phi(a^4), \phi(a^5), \phi(e)\} \\ &= \{a^3, a^6 = e, a^9 = a^3, a^{12} = e, a^{15} = a^3, e\} \\ &= \{a^3, e\} \\ &= H. \end{aligned}$$

Thus, G has H as a homomorphic image.

23. Let G be a group with center $\mathbf{Z}(G) = C$, and assume that G/C is cyclic, say $G/C = \langle gC \rangle$ for some $g \in G$. Let x and y be arbitrary elements of G. Since the left cosets of C in G form a partition of G (by Lemma 4.11), $x \in g^iC$ and $y \in g^jC$ for some powers i and j. That is, $x = g^ia$ and $y = g^jb$ for elements a and b in C. Then

$$\begin{aligned} xy &= g^iag^jb \\ &= g^ig^jab \quad \text{since } a \in C \\ &= g^{i+j}ab \quad \text{by Theorem 3.12b,} \end{aligned}$$

and

$$\begin{aligned} yx &= g^jbg^ia \\ &= g^jg^iba \quad \text{since } b \in C \\ &= g^{j+i}ba \quad \text{by Theorem 3.12b} \\ &= g^{i+j}ab \quad \text{since } j+i = i+j \text{ and } b \in C. \end{aligned}$$

Thus $xy = yx$, and G is an abelian group.

Exercises 4.6

6. $\mathbf{Z}_{20} = \langle [4] \rangle \oplus \langle [5] \rangle$ **10.** a. 2 b. 2 c. 6 d. 3

11. a. $\{([0],[0])\}$ $\{([0],[0]),([0],[1]),([0],[2]),([0],[3])\}$

$\{([0],[0]),([0],[2])\}$ $\{([0],[0]),([0],[2]),([1],[0]),([1],[2])\}$

$\{([0],[0]),([1],[0])\}$ $\{([0],[0]),([0],[2]),([1],[1]),([1],[3])\}$

$\{([0],[0]),([1],[2])\}$ $\mathbf{Z}_2 \oplus \mathbf{Z}_4$

b. $\{([0],[0])\}$
$\{([0],[0]),([0],[3])\}$
$\{([0],[0]),([1],[0])\}$
$\{([0],[0]),([1],[3])\}$
$\{([0],[0]),([0],[2]),([0],[4])\}$
$\{([0],[0]),([0],[3]),([1],[0]),([1],[3])\}$
$\{([0],[0]),([0],[2]),([0],[4]),([1],[0]),([1],[2]),([1],[4])\}$
$\{([0],[0]),([0],[1]),([0],[2]),([0],[3]),([0],[4]),([0],[5])\}$
$\{([0],[0]),([1],[1]),([0],[2]),([1],[3]),([0],[4]),([1],[5])\}$
$\mathbf{Z}_2 \oplus \mathbf{Z}_6$

12. a. $\mathbf{Z}_{15} = \langle [5] \rangle \oplus \langle [3] \rangle$, where $\langle [5] \rangle = \{[5],[10],[0]\}$ is a cyclic group of order 3 and $\langle [3] \rangle = \{[3],[6],[9],[12],[0]\}$ is a cyclic group of order 5. From this, it is intuitively clear that $\mathbf{Z}_{15}$ is isomorphic to $\mathbf{Z}_3 \oplus \mathbf{Z}_5$. The idea can be formalized as follows. For each $a \in \mathbf{Z}$, let $[a]_{15}, [a]_3$, and $[a]_5$ denote the congruence class of a modulo 15, 3, and 5, respectively. Any $[a]_{15}$ and $[b]_{15}$ in $\mathbf{Z}_{15}$ can be written as

$$[a]_{15} = r[5]_{15} + s[3]_{15} \quad \text{and} \quad [b]_{15} = p[5]_{15} + q[3]_{15}$$

with $r, s, p,$ and q integers. Since

$$\begin{aligned} [a]_{15} = [b]_{15} &\Leftrightarrow r[5]_{15} + s[3]_{15} = p[5]_{15} + q[3]_{15} \\ &\Leftrightarrow (r-p)[5]_{15} = (q-s)[3]_{15} \\ &\Leftrightarrow (r-p)[5]_{15} = [0]_{15} = (q-s)[3]_{15} \\ &\Leftrightarrow r - p \equiv 0 \pmod 3 \quad \text{and} \quad q - s \equiv 0 \pmod 5 \\ &\Leftrightarrow [r]_3 = [p]_3 \quad \text{and} \quad [q]_5 = [s]_5, \end{aligned}$$

the rule

$$\phi([a]_{15}) = ([r]_3, [s]_5)$$

defines a one-to-one mapping from $\mathbf{Z}_{15}$ to the external direct sum $\mathbf{Z}_3 \oplus \mathbf{Z}_5$. ϕ is clearly onto, and ϕ is a homomorphism since

$$\begin{aligned} \phi([a]_{15} + [b]_{15}) &= \phi((r+p)[5]_{15} + (s+q)[3]_{15}) \\ &= ([r+p]_3, [s+q]_5) \\ &= ([r]_3 + [p]_3, [s]_5 + [q]_5) \\ &= ([r]_3, [s]_5) + ([p]_3, [q]_5) \\ &= \phi([a]_{15}) + \phi([b]_{15}). \end{aligned}$$

Thus ϕ is an isomorphism from $\mathbf{Z}_{15}$ to $\mathbf{Z}_3 \oplus \mathbf{Z}_5$.

b. $\mathbf{Z}_{12} = \langle [4] \rangle \oplus \langle [3] \rangle$, where $\langle [4] \rangle = \{[4], [8], [0]\}$ is a cyclic group of order 3 and $\langle [3] \rangle = \{[3], [6], [9], [0]\}$ is a cyclic group of order 4. From this, it is intuitively clear that $\mathbf{Z}_{12}$ is isomorphic to $\mathbf{Z}_3 \oplus \mathbf{Z}_4$. The idea can be formalized as follows. For each $a \in \mathbf{Z}$, let $[a]_{12}, [a]_3$, and $[a]_4$ denote the congruence class of a modulo 12, 3, and 4, respectively. Any $[a]_{12}$ and $[b]_{12}$ in $\mathbf{Z}_{12}$ can be written as

$$[a]_{12} = r[4]_{12} + s[3]_{12} \quad \text{and} \quad [b]_{12} = p[4]_{12} + q[3]_{12}$$

with r, s, p, and q integers. Since

$$\begin{aligned} [a]_{12} = [b]_{12} &\Leftrightarrow r[4]_{12} + s[3]_{12} = p[4]_{12} + q[3]_{12} \\ &\Leftrightarrow (r-p)[4]_{12} = (q-s)[3]_{12} \\ &\Leftrightarrow (r-p)[4]_{12} = [0]_{12} = (q-s)[3]_{12} \\ &\Leftrightarrow r - p \equiv 0 \pmod 3 \quad \text{and} \quad q - s \equiv 0 \pmod 4 \\ &\Leftrightarrow [r]_3 = [p]_3 \quad \text{and} \quad [q]_4 = [s]_4, \end{aligned}$$

the rule

$$\phi([a]_{12}) = ([r]_3, [s]_4)$$

defines a one-to-one mapping from $\mathbf{Z}_{12}$ to the external direct sum $\mathbf{Z}_3 \oplus \mathbf{Z}_4$. ϕ is clearly onto, and ϕ is a homomorphism since

$$\begin{aligned} \phi([a]_{12} + [b]_{12}) &= \phi((r+p)[4]_{12} + (s+q)[3]_{12}) \\ &= ([r+p]_3, [s+q]_4) \\ &= ([r]_3 + [p]_3, [s]_4 + [q]_4) \\ &= ([r]_3, [s]_4) + ([p]_3, [q]_4) \\ &= \phi([a]_{12}) + \phi([b]_{12}). \end{aligned}$$

Thus ϕ is an isomorphism from $\mathbf{Z}_{12}$ to $\mathbf{Z}_3 \oplus \mathbf{Z}_4$.

Exercises 4.7

1. The cyclic group $C_9 = \langle a \rangle$ of order 9 is a $p-$ group with $p = 3$.

2. The permutation groups

$$G = \langle (1,2,3,4) \rangle \subseteq S_4 \quad \text{and} \quad G' = \{(1), (1,2), (3,4), (1,2)(3,4)\} \subseteq S_4$$

are 2-groups of order 4 that are not isomorphic.

3. a. $\langle (1,2,3) \rangle, \langle (1,2,4) \rangle, \langle (1,3,4) \rangle, \langle (2,3,4) \rangle$

   b. $\{(1), (1,2)(3,4), (1,3)(2,4), (1,4)(2,3)\}$ is the only Sylow 2-subgroup of A_4.

4. $\langle(1,2,3)\rangle$, $\langle(1,2,4)\rangle$, $\langle(1,3,4)\rangle$, $\langle(2,3,4)\rangle$

5. a. $$\begin{aligned}\mathbf{Z}_{10} &= \langle[5]\rangle \oplus \langle[2]\rangle \\ &= \{[0],[5]\} \oplus \{[0],[2],[4],[6],[8]\} \\ &= C_2 \oplus C_5\end{aligned}$$

b. $$\begin{aligned}\mathbf{Z}_{15} &= \langle[5]\rangle \oplus \langle[3]\rangle \\ &= \{[0],[5],[10]\} \oplus \{[0],[3],[6],[9],[12]\} \\ &= C_3 \oplus C_5\end{aligned}$$

c. $$\begin{aligned}\mathbf{Z}_{12} &= \langle[3]\rangle \oplus \langle[4]\rangle \\ &= \{[0],[3],[6],[9]\} \oplus \{[0],[4],[8]\} \\ &= C_4 \oplus C_3\end{aligned}$$

d. $$\begin{aligned}\mathbf{Z}_{18} &= \langle[9]\rangle \oplus \langle[2]\rangle \\ &= \{[0],[9]\} \oplus \{[0],[2],[4],[6],[8],[10],[12],[14],[16]\} \\ &= C_2 \oplus C_9\end{aligned}$$

6. a. Any abelian group of order 6 is isomorphic to $C_3 \oplus C_2$, where C_n is a cyclic group of order n.

b. Any abelian group of order 10 is isomorphic to $C_2 \oplus C_5$, where C_n is a cyclic group of order n.

c. Any abelian group of order 12 is isomorphic to either $C_4 \oplus C_3$ or $C_2 \oplus C_2 \oplus C_3$.

d. Any abelian group of order 18 is isomorphic to either $C_9 \oplus C_2$ or $C_3 \oplus C_3 \oplus C_2$.

e. Any abelian group of order 36 is isomorphic to one of the direct sums $C_4 \oplus C_9$, $C_2 \oplus C_2 \oplus C_9$, $C_4 \oplus C_3 \oplus C_3$, or $C_2 \oplus C_2 \oplus C_3 \oplus C_3$.

f. Any abelian group of order 100 is isomorphic to one of the direct sums

$$C_4 \oplus C_{25}, \quad C_2 \oplus C_2 \oplus C_{25}, \quad C_4 \oplus C_5 \oplus C_5, \quad \text{or} \quad C_2 \oplus C_2 \oplus C_5 \oplus C_5.$$

8. The permutation group $G = S_3 = \{(1), (1,2), (1,3), (2,3), (1,2,3), (1,3,2)\}$ is a nonabelian group with order that is divisible by the prime 2, but the set $\{(1), (1,2), (1,3), (2,3)\}$ of all elements that have orders that are powers of 2 is not a subgroup of G since it is not closed.

10. b. 16 11. b. There are 24 distinct elements of G that have order 6.

Exercises 5.1

2. a. Ring b. Ring

c. Not a ring. The set is not closed with respect to multiplication. For example, $\sqrt[3]{5}$ is in the set, but the product $\sqrt[3]{5} \cdot \sqrt[3]{5} = \sqrt[3]{25}$ is not in the set.

d. Ring

e. Not a ring. The set of positive real numbers does not contain an additive identity.

f. Ring g. Ring h. Ring

3.

$+$	$\varnothing$	A	B	U
$\varnothing$	$\varnothing$	A	B	U
A	A	$\varnothing$	U	B
B	B	U	$\varnothing$	A
U	U	B	A	$\varnothing$

$\cdot$	$\varnothing$	A	B	U
$\varnothing$	$\varnothing$	$\varnothing$	$\varnothing$	$\varnothing$
A	$\varnothing$	A	$\varnothing$	A
B	$\varnothing$	$\varnothing$	B	B
U	$\varnothing$	A	B	U

4. Let $U = \{a, b, c\}$, $A = \{a\}$, $B = \{b\}$, $C = \{c\}$, $A' = \{b, c\}$, $B' = \{a, c\}$, $C' = \{a, b\}$.

$+$	$\varnothing$	A	B	C	A'	B'	C'	U
$\varnothing$	$\varnothing$	A	B	C	A'	B'	C'	U
A	A	$\varnothing$	C'	B'	U	C	B	A'
B	B	C'	$\varnothing$	A'	C	U	A	B'
C	C	B'	A'	$\varnothing$	B	A	U	C'
A'	A'	U	C	B	$\varnothing$	C'	B'	A
B'	B'	C	U	A	C'	$\varnothing$	A'	B
C'	C'	B	A	U	B'	A'	$\varnothing$	C
U	U	A'	B'	C'	A	B	C	$\varnothing$

$\cdot$	$\varnothing$	A	B	C	A'	B'	C'	U
$\varnothing$	$\varnothing$	$\varnothing$	$\varnothing$	$\varnothing$	$\varnothing$	$\varnothing$	$\varnothing$	$\varnothing$
A	$\varnothing$	A	$\varnothing$	$\varnothing$	$\varnothing$	A	A	A
B	$\varnothing$	$\varnothing$	B	$\varnothing$	B	$\varnothing$	B	B
C	$\varnothing$	$\varnothing$	$\varnothing$	C	C	C	$\varnothing$	C
A'	$\varnothing$	$\varnothing$	B	C	A'	C	B	A'
B'	$\varnothing$	A	$\varnothing$	C	C	B'	A	B'
C'	$\varnothing$	A	B	$\varnothing$	B	A	C'	C'
U	$\varnothing$	A	B	C	A'	B'	C'	U

5. The set $\mathcal{P}(A)$ is not a ring with respect to the operations of addition and multiplication as defined, since the set does not contain additive inverse elements.

6. $\mathcal{P}(U)$ is not a ring with respect to the given operations. Condition 4 of Definition 5.1a fails because the element U does not have an additive inverse.

7. **a**. $[2], [3], [4]$ **b**. $[2], [4], [6]$ **c**. $[2], [4], [5], [6], [8]$

d. $[2], [3], [4], [6], [8], [9], [10]$ **e**. $[2], [4], [6], [7], [8], [10], [12]$

f. If n is a prime integer, there are no zero divisors in $\mathbf{Z}_n$.

8. **a**. $[1]^{-1} = [1], [5]^{-1} = [5]$ **b**. Each of $[1], [3], [5]$, and $[7]$ is its own inverse.

c. $[1]^{-1} = [1], [3]^{-1} = [11], [5]^{-1} = [13], [7]^{-1} = [7], [9]^{-1} = [9],$
$[11]^{-1} = [3], [13]^{-1} = [5], [15]^{-1} = [15]$

d. Each of $[1], [5], [7]$, and $[11]$ is its own inverse.

e. $[1]^{-1} = [1], [3]^{-1} = [5], [5]^{-1} = [3], [9]^{-1} = [11], [11]^{-1} = [9], [13]^{-1} = [13]$

f. If n is a prime integer, all nonzero elements of $\mathbf{Z}_n$ have multiplicative inverses.

17. Let $R_1 = \mathbf{E}$ and R_2 be the set of all multiples of five. Each of R_1 and R_2 are subrings of $\mathbf{Z}$, but $R_1 \cup R_2$ is not, since it is not closed with respect to addition.

18. In the ring $M_2(\mathbf{Z})$, let $a = \begin{bmatrix} 1 & 0 \\ 0 & 0 \end{bmatrix}$ and $b = \begin{bmatrix} 0 & 0 \\ 1 & 0 \end{bmatrix}$. Then $ab = \begin{bmatrix} 0 & 0 \\ 0 & 0 \end{bmatrix}$

but $ba = \begin{bmatrix} 0 & 0 \\ 1 & 0 \end{bmatrix} \neq \begin{bmatrix} 0 & 0 \\ 0 & 0 \end{bmatrix}$.

24. **a.**

$+$	$[0]$	$[2]$	$[4]$	$[6]$	$[8]$
$[0]$	$[0]$	$[2]$	$[4]$	$[6]$	$[8]$
$[2]$	$[2]$	$[4]$	$[6]$	$[8]$	$[0]$
$[4]$	$[4]$	$[6]$	$[8]$	$[0]$	$[2]$
$[6]$	$[6]$	$[8]$	$[0]$	$[2]$	$[4]$
$[8]$	$[8]$	$[0]$	$[2]$	$[4]$	$[6]$

$\times$	$[0]$	$[2]$	$[4]$	$[6]$	$[8]$
$[0]$	$[0]$	$[0]$	$[0]$	$[0]$	$[0]$
$[2]$	$[0]$	$[4]$	$[8]$	$[2]$	$[6]$
$[4]$	$[0]$	$[8]$	$[6]$	$[4]$	$[2]$
$[6]$	$[0]$	$[2]$	$[4]$	$[6]$	$[8]$
$[8]$	$[0]$	$[6]$	$[2]$	$[8]$	$[4]$

c. R is a subring of $\mathbf{Z}_{10}$. **d.** No **e.** All except $[0]$.

25. **a.** Yes **b.** The set S is a commutative ring, and it contains the unity $[10]$.

c. Yes **d.** Yes, $[6]$ and $[12]$ **e.** $[2], [4], [8], [10], [14], [16]$

26.

$\cdot$	a	b	c
a	a	a	a
b	a	c	b
c	a	b	c

27.

$\cdot$	a	b	c	d
a	a	a	a	a
b	a	c	a	c
c	a	a	a	a
d	a	c	a	c

28. $\begin{bmatrix} 1 & 1 \\ 0 & 0 \end{bmatrix}$ is a zero divisor since $\begin{bmatrix} 1 & 1 \\ 0 & 0 \end{bmatrix}\begin{bmatrix} 2 & 2 \\ -2 & -2 \end{bmatrix} = \begin{bmatrix} 0 & 0 \\ 0 & 0 \end{bmatrix}$.

29. $\begin{bmatrix} 1 & 0 \\ 0 & 1 \end{bmatrix}, \begin{bmatrix} 1 & 0 \\ 0 & 0 \end{bmatrix}$

31. **a.** S is a subring of $M_2(\mathbf{Z})$. **b.** S is a subring of $M_2(\mathbf{Z})$.

c. S is a subring of $M_2(\mathbf{Z})$. **d.** S is a subring of $M_2(\mathbf{Z})$.

32. **b.** All $\begin{bmatrix} a & b \\ 0 & c \end{bmatrix}$ with $a = \pm 1$ and $c = \pm 1$.

39. $\mathbf{Z}_2 \oplus \mathbf{Z}_2 = \{(0,0), (0,1), (1,0), (1,1)\}$

$+$	$(0,0)$	$(0,1)$	$(1,0)$	$(1,1)$
$(0,0)$	$(0,0)$	$(0,1)$	$(1,0)$	$(1,1)$
$(0,1)$	$(0,1)$	$(0,0)$	$(1,1)$	$(1,0)$
$(1,0)$	$(1,0)$	$(1,1)$	$(0,0)$	$(0,1)$
$(1,1)$	$(1,1)$	$(1,0)$	$(0,1)$	$(0,0)$

$\cdot$	$(0,0)$	$(0,1)$	$(1,0)$	$(1,1)$
$(0,0)$	$(0,0)$	$(0,0)$	$(0,0)$	$(0,0)$
$(0,1)$	$(0,0)$	$(0,1)$	$(0,0)$	$(0,1)$
$(1,0)$	$(0,0)$	$(0,0)$	$(1,0)$	$(1,0)$
$(1,1)$	$(0,0)$	$(0,1)$	$(1,0)$	$(1,1)$

Exercises 5.2

1. a. The set of real numbers of the form $m + n\sqrt{2}$ where m and n are integers is an integral domain. It is not a field, since not every element (for example, $2 + 0\sqrt{2}$) has a multiplicative inverse.

   b. The set is both an integral domain and a field.

   c. The set of real numbers of the form $a + b\sqrt[3]{2}$ where a and b are rational numbers is neither an integral domain nor a field, since it is not a ring. The set is not closed with respect to multiplication. For example: $\sqrt[3]{2} \cdot \sqrt[3]{2} = \sqrt[3]{4}$ is not in the set.

   d. The set is both an integral domain and a field.

   e. The set of all complex numbers of the form $m + ni$ where $m \in \mathbf{Z}$ and $n \in \mathbf{Z}$ is an integral domain. It is not a field, since not every element (for example, $2 + 0i$) has a multiplicative inverse.

   f. The set is neither an integral domain nor a field since it contains no unity element.

   g. The set of all complex numbers of the form $a + bi$ where a and b are rational numbers is both an integral domain and a field.

   h. The set is an integral domain. It is not a field since some elements (such as $3 + 0\sqrt{2}$) do not have multiplicative inverses in the set.

2. a. R is an integral domain. b. R is a field.

3. a. The set S is not an integral domain, since the elements $[6]$ and $[12]$ are zero divisors.

   b. The set S is not a field, since $[6]$ and $[12]$ do not have multiplicative inverses.

4. b. Yes c. Yes, $(1,1)$ is a unity. d. Yes e. Yes

5. The ring W is commutative, since if (x,y) and (z,w) are elements of W, we have

$$\begin{aligned} (x,y)\cdot(z,w) &= (xz - yw, xw + yz) \\ &= (zx - wy, zy + wx) \\ &= (z,w)\cdot(x,y). \end{aligned}$$

The element $(1,0)$ in W is the unity element, since for (x,y) in W we have

$$\begin{aligned}(x,y)\cdot(1,0) &= (1,0)\cdot(x,y)\\ &= (1x-0y, 1y+0x)\\ &= (x,y).\end{aligned}$$

6. **a**. Yes **b**. $\begin{bmatrix}1 & 0\\ 1 & 0\end{bmatrix}$ is a unity. **c**. Yes **d**. Yes

7. **a**. Yes **b**. $\begin{bmatrix}1 & 1\\ 0 & 0\end{bmatrix}$ is a unity. **c**. Yes **d**. Yes

8. The ring R is commutative with unity element $\begin{bmatrix}1 & 0\\ 0 & 1\end{bmatrix}$.

9. **a**. Yes **b**. $\begin{bmatrix}1 & 0\\ 0 & 1\end{bmatrix}$ is a unity. **c**. Yes **d**. Yes

14. **a**. Consider the ring $\mathbf{Z}_{10}$. The elements $[1]$ and $[3]$ are not zero divisors, but the sum $[1]+[3]=[4]$ is a zero divisor.

15. **a**. $[173]$ **b**. $[113]$ **c**. $[27]$ **d**. $[266]$

Exercises 5.3

9. Define $\phi : W \to R$ by

$$\phi((x,y)) = \begin{bmatrix}x & -y\\ y & x\end{bmatrix}.$$

The mapping ϕ is clearly a one-to-one correspondence from W to R.

$$\begin{aligned}\phi((x,y)+(z,w)) &= \phi((x+z, y+w))\\ &= \begin{bmatrix}x+z & -y-w\\ y+w & x+z\end{bmatrix}\\ &= \begin{bmatrix}x & -y\\ y & x\end{bmatrix} + \begin{bmatrix}z & -w\\ w & z\end{bmatrix}\\ &= \phi((x,y)) + \phi((z,w))\end{aligned}$$

$$\begin{aligned}
\phi((x,y)\cdot(z,w)) &= \phi((xz-yw, xw+yz)) \\
&= \begin{bmatrix} xz-yw & -xw-yz \\ xw+yz & xz-yw \end{bmatrix} \\
&= \begin{bmatrix} x & -y \\ y & x \end{bmatrix} \cdot \begin{bmatrix} z & -w \\ w & z \end{bmatrix} \\
&= \phi((x,y)) \cdot \phi((z,w))
\end{aligned}$$

Thus, ϕ is an isomorphism.

10. The mapping $\phi : \mathcal{C} \to R$ defined by $\phi(a+bi) = \begin{bmatrix} a & -b \\ b & a \end{bmatrix}$ is an isomorphism from $\mathcal{C}$ to R.

11. **a.** For notational convenience in this solution, we write 0 for $[0]$, 1 for $[1]$, and 2 and $[2]$ in $\mathbf{Z}_3$. Then

$$S = \{(0,1), (0,2), (1,1), (1,2), (2,1), (2,2)\}.$$

Since $(0,1) \sim (0,2), (1,1) \sim (2,2)$ and $(1,2) \sim (2,1)$ in S, the distinct elements of Q are $[0,1], [1,1]$, and $[2,1]$.

b. Define $\phi : D \to Q$ by

$$\begin{aligned}
\phi(0) &= [0,1] \\
\phi(1) &= [1,1] \\
\phi(2) &= [2,1].
\end{aligned}$$

12. **a.** For notational convenience, we write a for $[a]$ in $\mathbf{Z}_5 = \{[0], [1], [2], [3], [4]\}$, and $S = \{(0,1), (0,2), (0,3), (0,4), (1,1), (1,2), (1,3), (1,4), (2,1), (2,2), (2,3), (2,4), (3,1), (3,2), (3,3), (3,4), (4,1), (4,2), (4,3), (4,4)\}$. We have

$$\begin{aligned}
&(0,1) \sim (0,2) \sim (0,3) \sim (0,4) \\
&(1,1) \sim (2,2) \sim (3,3) \sim (4,4) \\
&(2,1) \sim (4,2) \sim (1,3) \sim (3,4) \\
&(3,1) \sim (1,2) \sim (4,3) \sim (2,4) \\
&(4,1) \sim (3,2) \sim (2,3) \sim (1,4)
\end{aligned}$$

so the distinct elements of Q are

$$[0,1], \quad [1,1], \quad [2,1], \quad [3,1], \quad \text{and} \quad [4,1].$$

b. An isomorphism ϕ from D to Q is defined by $\phi(x) = [x,1]$ for all x in D.

15. The set of all quotients for D is the set Q of all equivalence classes $[m+ni, r+si]$, where $m+ni \in D$ and $r+si \in D$ with not both r and s equal 0. To show that Q is isomorphic to the set C of all complex numbers of the form $a+bi$ where a and b are rational numbers, we define $\phi : Q \to C$ by

$$\phi([m+ni, r+si]) = \frac{m+ni}{r+si}.$$

This rule defines a mapping from Q into C, since for $[m+ni, r+si] \in Q$ we can write

$$\frac{m+ni}{r+si} = \frac{mr+ns}{r^2+s^2} + \frac{nr-ms}{r^2+s^2}i,$$

which is an element in C.

To show that ϕ is onto, let $a+bi$ be an arbitrary element in C. Since a and b are both rational numbers, there exist integers $p, q, t,$ and u such that

$$a = \frac{p}{q} \quad \text{and} \quad b = \frac{t}{u}.$$

Then the element $[pu+qti, qu+0i]$ is in Q, and

$$\begin{aligned} \phi([pu+qti, qu+0i]) &= \frac{pu+qti}{qu+0i} \\ &= \frac{p}{q} + \frac{t}{u}i \\ &= a+bi. \end{aligned}$$

To show that ϕ is one-to-one, let $[m+ni, r+si]$ and $[x+yi, z+wi]$ be elements of Q such that

$$\phi([m+ni, r+si]) = \phi([x+yi, z+wi]).$$

Then

$$\frac{m+ni}{r+si} = \frac{x+yi}{z+wi},$$

and this implies that

$$(m+ni)(z+wi) = (r+si)(x+yi).$$

By the definition of equality in Q, we have

$$[m+ni, r+si] = [x+yi, z+wi],$$

and therefore ϕ is one-to-one. Since

$$\begin{aligned} &\phi([m+ni, r+si] + [x+yi, z+wi]) \\ &= \phi([(m+ni)(z+wi) + (r+si)(x+yi), (r+si)(z+wi)]) \\ &= \frac{(m+ni)(z+wi) + (r+si)(x+yi)}{(r+si)(z+wi)} \\ &= \frac{m+ni}{r+si} + \frac{x+yi}{z+wi} \\ &= \phi([m+ni, r+si]) + \phi([x+yi, z+wi]) \end{aligned}$$

and

$$\begin{aligned}
&\phi([m+ni, r+si]\cdot[x+yi, z+wi]) \\
&= \phi([(m+ni)(x+yi), (r+si)(z+wi)]) \\
&= \frac{(m+ni)(x+yi)}{(r+si)(z+wi)} \\
&= \frac{m+ni}{r+si}\cdot\frac{x+yi}{z+wi} \\
&= \phi([m+ni, r+si])\cdot\phi([x+yi, z+wi]),
\end{aligned}$$

ϕ is an isomorphism from Q to C.

18. $\frac{m}{2^n}$ for $m, n \in \mathbf{Z}$

Exercises 6.1

5. a. Let I_1 and I_2 be the principal ideals $I_1 = (2)$ and $I_2 = (3)$ of $\mathbf{Z}$. Then 2 and 3 are in $I_1 \cup I_2$, but the sum $2+3=5$ is not in $I_1 \cup I_2$. Hence $I_1 \cup I_2$ is not an ideal of $\mathbf{Z}$.

   b. Let I_1 and I_2 be the principal ideals $I_1 = (2)$ and $I_2 = (4)$ of $\mathbf{Z}$. Then $I_1 \cup I_2 = (2) = I_1$ is an ideal of $\mathbf{Z}$.

13. a. $\{[0]\}, \mathbf{Z}_7$ b. $\{[0]\}, \mathbf{Z}_{11}$ c. $\{[0]\}$

$([6]) = \{[0], [6]\}$

$([4]) = \{[0], [4], [8]\}$

$([3]) = \{[0], [3], [6], [9]\}$

$([2]) = \{[0], [2], [4], [6], [8], [10]\}$

$\mathbf{Z}_{12}$

d. $\{[0]\}$

$([9]) = \{[0], [9]\}$

$([6]) = \{[0], [6], [12]\}$

$([3]) = \{[0], [3], [6], [9], [12], [15]\}$

$([2]) = \{[0], [2], [4], [6], [8], [10], [12], [14], [16]\}$

$\mathbf{Z}_{18}$

e. $\{[0]\}$

$([10]) = \{[0], [10]\}$

$([5]) = \{[0], [5], [10], [15]\}$

$([4]) = \{[0], [4], [8], [12], [16]\}$

$([2]) = \{[0], [2], [4], [6], [8], [10], [12], [14], [16], [18]\}$

$\mathbf{Z}_{20}$

f. $\{[0]\}$

$([12]) = \{[0], [12]\}$

$([8]) = \{[0], [8], [16]\}$

$([6]) = \{[0], [6], [12], [18]\}$

$([4]) = \{[0], [4], [8], [12], [16], [20]\}$

$([3]) = \{[0], [3], [6], [9], [12], [15], [18], [21]\}$

$([2]) = \{[0], [2], [4], [6], [8], [10], [12], [14], [16], [18], [20], [22]\}$

$\mathbf{Z}_{24}$

17. The set U is not an ideal of S. $X = \begin{bmatrix} 1 & 2 \\ 0 & 1 \end{bmatrix}$ is in U, and $R = \begin{bmatrix} 1 & 2 \\ 0 & 3 \end{bmatrix}$ is in S, but $XR = \begin{bmatrix} 1 & 8 \\ 0 & 3 \end{bmatrix}$ is not in U.

18. **b**. No **c**. No

d. U is an ideal of R. It is clearly nonempty. Since

$$\begin{bmatrix} 0 & 0 \\ x & 0 \end{bmatrix} + \begin{bmatrix} 0 & 0 \\ y & 0 \end{bmatrix} = \begin{bmatrix} 0 & 0 \\ x+y & 0 \end{bmatrix},$$

U is closed under addition. And since

$$-\begin{bmatrix} 0 & 0 \\ a & 0 \end{bmatrix} = \begin{bmatrix} 0 & 0 \\ -a & 0 \end{bmatrix},$$

U contains additive inverses for its elements. For arbitrary $\begin{bmatrix} 0 & 0 \\ a & 0 \end{bmatrix}$ in U

and $\begin{bmatrix} x & 0 \\ y & 0 \end{bmatrix}$ in R, the products

$$\begin{bmatrix} 0 & 0 \\ a & 0 \end{bmatrix}\begin{bmatrix} x & 0 \\ y & 0 \end{bmatrix} = \begin{bmatrix} 0 & 0 \\ ax & 0 \end{bmatrix} \text{ and } \begin{bmatrix} x & 0 \\ y & 0 \end{bmatrix}\begin{bmatrix} 0 & 0 \\ a & 0 \end{bmatrix} = \begin{bmatrix} 0 & 0 \\ 0 & 0 \end{bmatrix}$$

are in U.

Exercises 6.2

4. b. $\ker\theta = \left\{ \begin{bmatrix} x & y \\ 0 & 0 \end{bmatrix} \middle| \, x \in \mathbf{Z} \text{ and } y \in \mathbf{Z} \right\}$,

$$\phi\left(\begin{bmatrix} x & y \\ 0 & z \end{bmatrix} + \ker\theta\right) = \theta\left(\begin{bmatrix} x & y \\ 0 & z \end{bmatrix}\right) = z$$

5. b. $\ker\theta = \left\{ \begin{bmatrix} 0 & 0 \\ y & 0 \end{bmatrix} \middle| \, y \in \mathbf{Z} \right\}$,

$$\phi\left(\begin{bmatrix} x & 0 \\ y & 0 \end{bmatrix} + \ker\theta\right) = \theta\left(\begin{bmatrix} x & 0 \\ y & 0 \end{bmatrix}\right) = x$$

6. b. $\ker\theta = \{[0]_6, [2]_6, [4]_6\}$ **9.** $\ker\theta = \left\{ \begin{bmatrix} 2m & 2n \\ 2p & 2q \end{bmatrix} \middle| \, m, n, p, q \in \mathbf{Z} \right\}$

11. θ is a homomorphism, since

$$\begin{aligned} \theta([a]+[b]) &= 4([a]+[b]) \\ &= 4[a]+4[b] \\ &= \theta([a])+\theta([b]) \end{aligned} \qquad \text{and} \qquad \begin{aligned} \theta([a])\,\theta([b]) &= (4[a])(4[b]) \\ &= 16[ab] \\ &= [16ab] \\ &= [4ab] \\ &= \theta([ab]) \\ &= \theta([a][b]). \end{aligned}$$

12. The mapping $\phi : R \to \mathbf{Z}_3$ given by

$$\phi(a) = [0], \quad \phi(b) = [2], \quad \phi(c) = [1]$$

is an isomorphism.

13.

+	[0]	[2]	[4]	[6]
[0]	[0]	[2]	[4]	[6]
[2]	[2]	[4]	[6]	[0]
[4]	[4]	[6]	[0]	[2]
[6]	[6]	[0]	[2]	[4]

·	[0]	[2]	[4]	[6]
[0]	[0]	[0]	[0]	[0]
[2]	[0]	[4]	[0]	[4]
[4]	[0]	[0]	[0]	[0]
[6]	[0]	[4]	[0]	[4]

The mapping $\phi : R \to R'$ given by

$$\phi(a) = [0], \quad \phi(b) = [2], \quad \phi(c) = [4], \quad \phi(d) = [6]$$

is an isomorphism.

15. a. θ does not preserve addition, θ preserves multiplication, θ is not a homomorphism.

b. θ preserves addition, θ does not preserve multiplication, θ is not a homomorphism.

c. θ preserves addition, θ does not preserve multiplication, θ is not a homomorphism.

d. θ does not preserve addition, θ preserves multiplication, θ is not a homomorphism.

e. θ does not preserve addition, θ preserves multiplication, θ is not a homomorphism.

f. θ does not preserve addition, θ does not preserve multiplication, θ is not a homomorphism.

16. a. The ideals of $\mathbf{Z}_6$ are

$$I_1 = \{[0]\}, \qquad I_3 = \{[0], [2], [4]\},$$
$$I_2 = \{[0], [3]\}, \qquad I_4 = \mathbf{Z}_6.$$

The quotient rings are as follows.

(1) $\mathbf{Z}_6/I_1$ is isomorphic to $\mathbf{Z}_6$.
(2) $\mathbf{Z}_6/I_2 = \{I_2,\ [1] + I_2,\ [2] + I_2\}$ is isomorphic to $\mathbf{Z}_3$.
(3) $\mathbf{Z}_6/I_3 = \{I_3,\ [1] + I_3\}$ is isomorphic to $\mathbf{Z}_2$.
(4) $\mathbf{Z}_6/\mathbf{Z}_6 = \{\mathbf{Z}_6\}$ is a ring with only the zero element.

The homomorphic images of $\mathbf{Z}_6$ are (isomorphic to) $\mathbf{Z}_6, \mathbf{Z}_3, \mathbf{Z}_2$, and $\{0\}$.

b. The ideals of $\mathbf{Z}_{10}$ are

$$I_1 = \{[0]\}, \qquad I_3 = \{[0], [2], [4], [6], [8]\},$$
$$I_2 = \{[0], [5]\}, \qquad I_4 = \mathbf{Z}_{10}.$$

The quotient rings are as follows.

(1) $\mathbf{Z}_{10}/I_1$ is isomorphic to $\mathbf{Z}_{10}$.

(2) $\mathbf{Z}_{10}/I_2 = \{I_2,\ [1]+I_2,\ [2]+I_2,\ [3]+I_2,\ [4]+I_2\}$ is isomorphic to $\mathbf{Z}_5$.

(3) $\mathbf{Z}_{10}/I_3 = \{I_3,\ [1]+I_3\}$ is isomorphic to $\mathbf{Z}_2$.

(4) $\mathbf{Z}_{10}/\mathbf{Z}_{10} = \{\mathbf{Z}_{10}\}$ is a ring with only the zero element.

The homomorphic images of $\mathbf{Z}_{10}$ are (isomorphic to) $\mathbf{Z}_{10}, \mathbf{Z}_5, \mathbf{Z}_2$, and $\{0\}$.

c. The ideals of $\mathbf{Z}_{12}$ are

$$I_1 = \{[0]\}, \qquad I_4 = \{[0], [3], [6], [9]\},$$
$$I_2 = \{[0], [6]\}, \qquad I_5 = \{[0], [2], [4], [6], [8], [10]\},$$
$$I_3 = \{[0], [4], [8]\}, \qquad I_6 = \mathbf{Z}_{12}.$$

The quotient rings are as follows.

(1) $\mathbf{Z}_{12}/I_1$ is isomorphic to $\mathbf{Z}_{12}$.

(2) $\mathbf{Z}_{12}/I_2 = \{I_2,\ [1]+I_2,\ [2]+I_2,\ [3]+I_2,\ [4]+I_2,\ [5]+I_2\}$ is isomorphic to $\mathbf{Z}_6$.

(3) $\mathbf{Z}_{12}/I_3 = \{I_3,\ [1]+I_3,\ [2]+I_3,\ [3]+I_3\}$ is isomorphic to $\mathbf{Z}_4$.

(4) $\mathbf{Z}_{12}/I_4 = \{I_4,\ [1]+I_4,\ [2]+I_4\}$ is isomorphic to $\mathbf{Z}_3$.

(5) $\mathbf{Z}_{12}/I_5 = \{I_5,\ [1]+I_5\}$ is isomorphic to $\mathbf{Z}_2$.

(6) $\mathbf{Z}_{12}/\mathbf{Z}_{12} = \{\mathbf{Z}_{12}\}$ is a ring with only the zero element.

The homomorphic images of $\mathbf{Z}_{12}$ are (isomorphic to) $\mathbf{Z}_{12}, \mathbf{Z}_6, \mathbf{Z}_4, \mathbf{Z}_3, \mathbf{Z}_2$, and $\{0\}$.

d. The ideals of $\mathbf{Z}_{18}$ are

$$I_1 = \{[0]\}, \qquad I_4 = \{[0], [3], [6], [9], [12], [15]\},$$
$$I_2 = \{[0], [9]\}, \qquad I_5 = \{[0], [2], [4], [6], [8], [10], [12], [14], [16]\},$$
$$I_3 = \{[0], [6], [12]\}, \qquad I_6 = \mathbf{Z}_{18}.$$

The quotient rings are as follows.

(1) $\mathbf{Z}_{18}/I_1$ is isomorphic to $\mathbf{Z}_{18}$.

(2) $\mathbf{Z}_{18}/I_2$ is isomorphic to $\mathbf{Z}_9$.

(3) $\mathbf{Z}_{18}/I_3$ is isomorphic to $\mathbf{Z}_6$.

(4) $\mathbf{Z}_{18}/I_4$ is isomorphic to $\mathbf{Z}_3$.

(5) $\mathbf{Z}_{18}/I_5$ is isomorphic to $\mathbf{Z}_2$.
(6) $\mathbf{Z}_{18}/I_6$ is isomorphic to $\{0\}$.

The homomorphic images of $\mathbf{Z}_{18}$ are (isomorphic to) $\mathbf{Z}_{18}, \mathbf{Z}_9, \mathbf{Z}_6, \mathbf{Z}_3, \mathbf{Z}_2$, and $\{0\}$.

e. The ideals of $\mathbf{Z}_8$ are

$$I_1 = \{[0]\}, \qquad I_3 = \{[0], [2], [4], [6]\},$$
$$I_2 = \{[0], [4]\}, \qquad I_4 = \mathbf{Z}_8.$$

The quotient rings are as follows.

(1) $\mathbf{Z}_8/I_1$ is isomorphic to $\mathbf{Z}_8$.
(2) $\mathbf{Z}_8/I_2 = \{I_2,\ [1]+I_2,\ [2]+I_2,\ [3]+I_2\}$ is isomorphic to $\mathbf{Z}_4$.
(3) $\mathbf{Z}_8/I_3 = \{I_3,\ [1]+I_3\}$ is isomorphic to $\mathbf{Z}_2$.
(4) $\mathbf{Z}_8/\mathbf{Z}_8 = \{\mathbf{Z}_8\}$ is a ring with only the zero element.

The homomorphic images of $\mathbf{Z}_8$ are (isomorphic to) $\mathbf{Z}_8, \mathbf{Z}_4, \mathbf{Z}_2$, and $\{0\}$.

f. The ideals of $\mathbf{Z}_{20}$ are

$$I_1 = \{[0]\}, \qquad I_4 = \{[0], [4], [8], [12], [16]\},$$
$$I_2 = \{[0], [10]\}, \qquad I_5 = \{[0], [2], [4], [6], [8], [10], [12], [14], [16], [18]\},$$
$$I_3 = \{[0], [5], [10], [15]\}, \quad I_6 = \mathbf{Z}_{20}.$$

The quotient rings are as follows.

(1) $\mathbf{Z}_{20}/I_1$ is isomorphic to $\mathbf{Z}_{20}$.
(2) $\mathbf{Z}_{20}/I_2$ is isomorphic to $\mathbf{Z}_{10}$.
(3) $\mathbf{Z}_{20}/I_3$ is isomorphic to $\mathbf{Z}_5$.
(4) $\mathbf{Z}_{20}/I_4$ is isomorphic to $\mathbf{Z}_4$.
(5) $\mathbf{Z}_{20}/I_5$ is isomorphic to $\mathbf{Z}_2$.
(6) $\mathbf{Z}_{20}/I_6$ is isomorphic to $\{0\}$.

The homomorphic images of $\mathbf{Z}_{20}$ are (isomorphic to) $\mathbf{Z}_{20}, \mathbf{Z}_{10}, \mathbf{Z}_5, \mathbf{Z}_4, \mathbf{Z}_2$, and $\{0\}$.

19. b. -1 is the zero element, and 0 is the unity of R'.

Exercises 6.3

1. a. 2 b. 3 c. 6 d. 4 e. 12

3. In $\mathbf{Z}_6$, $[1]$ has additive order 6, and $[2]$ has additive order 3.

7. **b.** Exercise 2 assures us that e, a, and b all have additive order 2. The other entries in the table can be determined by using the fact that D forms a group with respect to addition. For example, $e + a = a$ would imply $e = 0$, so $e + a = b$ must be true.

$+$	0	e	a	b
0	0	e	a	b
e	e	0	b	a
a	a	b	0	e
b	b	a	e	0

Exercises 6.4

5. $R/I = \{I, 1+I, \sqrt{2}+I, 1+\sqrt{2}+I\}$ **6.** $R/I = \{I, 1+I\}$

7. $\mathbf{E}/I = \{I, 2+I, 4+I\}$ **8.** $\mathbf{E}/I = \{I, 2+I, 4+I, 6+I, 8+I\}$

9. $\{[0], [3], [6], [9]\}$, $\{[0], [2], [4], [6], [8], [10]\}$

10. $\{[0], [3], [6], [9], [12], [15]\}$, $\{[0], [2], [4], [6], [8], [10], [12], [14], [16]\}$

19. $\{[0], [3], [6], [9]\}$, $\{[0], [2], [4], [6], [8], [10]\}$

20. $\{[0], [3], [6], [9], [12], [15]\}$, $\{[0], [2], [4], [6], [8], [10], [12], [14], [16]\}$

Exercises 7.1

1. $0.\overline{5}$ **2.** $0.\overline{21}$ **3.** $0.\overline{987654320}$ **4.** $2.\overline{285714}$ **5.** $3.\overline{142857}$ **6.** $1.\overline{72}$

7. $\frac{31}{9}$ **8.** $\frac{5}{3}$ **9.** $\frac{4}{33}$ **10.** $\frac{7}{11}$ **11.** $\frac{83}{33}$ **12.** $\frac{1070}{333}$

20. **a.** $a = \sqrt{2}$ and $b = -\sqrt{2}$ are irrational, but $a + b = 0$ is rational.

b. $a = \sqrt{2}$ and $b = -\sqrt{2}$ are irrational, but $ab = -2$ is rational.

21. **a.** An element v of F is a *lower bound* of S if $v \leq x$ for all $x \in S$. An element v of F is a *greatest lower bound* of S if these conditions are satisfied:

(1) v is a lower bound of S.

(2) If $b \in F$ is a lower bound of S, then $b \leq v$.

Exercises 7.2

1. $10 + 11i$ **2.** $-2 - 26i$ **3.** $-i$ **4.** $-i$ **5.** $2 - 11i$ **6.** $-4 + 3i$

7. $\frac{2}{5} + \frac{1}{5}i$ **8.** $\frac{3}{10} - \frac{1}{10}i$ **9.** $\frac{11}{50} + \frac{1}{25}i$ **10.** $-\frac{1}{5} - \frac{2}{5}i$ **11.** $\frac{21}{29} + \frac{20}{29}i$

12. $\frac{7}{25} - \frac{24}{25}i$

13. a. $3i, -3i$ b. $4i, -4i$ c. $5i, -5i$ d. $6i, -6i$ e. $\sqrt{13}\,i, -\sqrt{13}\,i$
f. $2\sqrt{2}\,i,\ -2\sqrt{2}\,i$

Exercises 7.3

1. a. $-2+2\sqrt{3}\,i = 4\left(\cos\frac{2\pi}{3} + i\sin\frac{2\pi}{3}\right)$

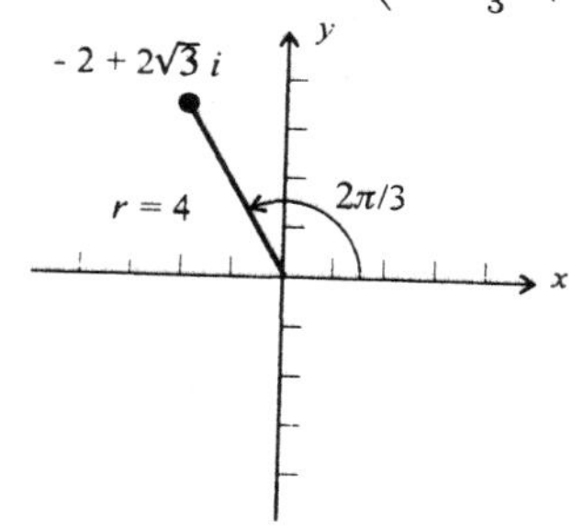

b. $2+2i = 2\sqrt{2}\left(\cos\frac{\pi}{4} + i\sin\frac{\pi}{4}\right)$

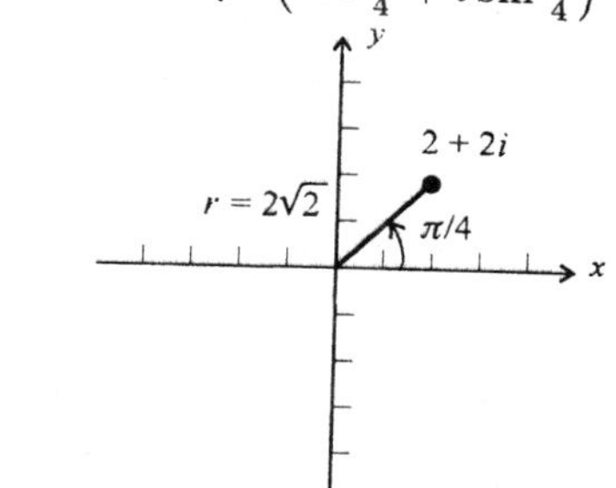

c. $3-3i = 3\sqrt{2}\left(\cos\frac{7\pi}{4} + i\sin\frac{7\pi}{4}\right)$

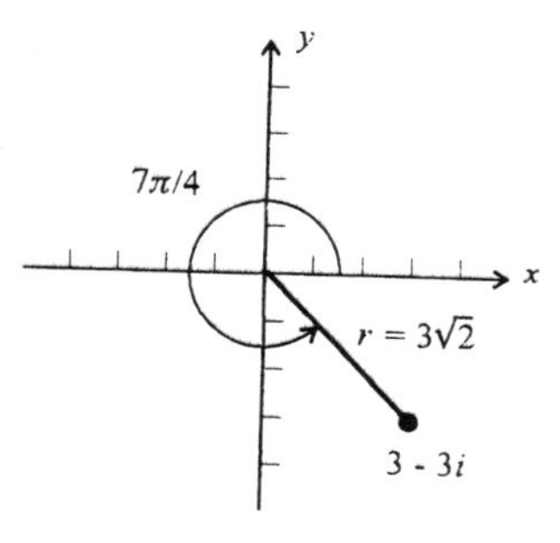

d. $\sqrt{3}+i = 2\left(\cos\frac{\pi}{6} + i\sin\frac{\pi}{6}\right)$

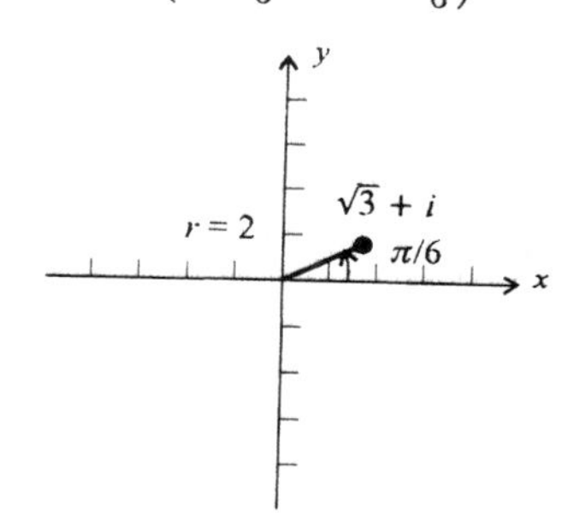

e. $1+\sqrt{3}\,i = 2\left(\cos\frac{\pi}{3} + i\sin\frac{\pi}{3}\right)$

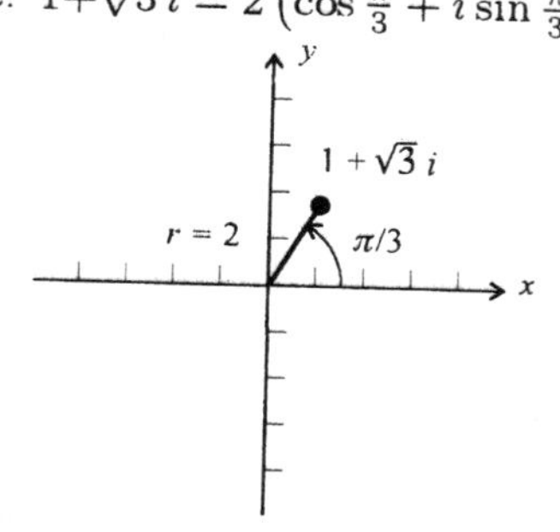

f. $-1-i = \sqrt{2}\left(\cos\frac{5\pi}{4} + i\sin\frac{5\pi}{4}\right)$

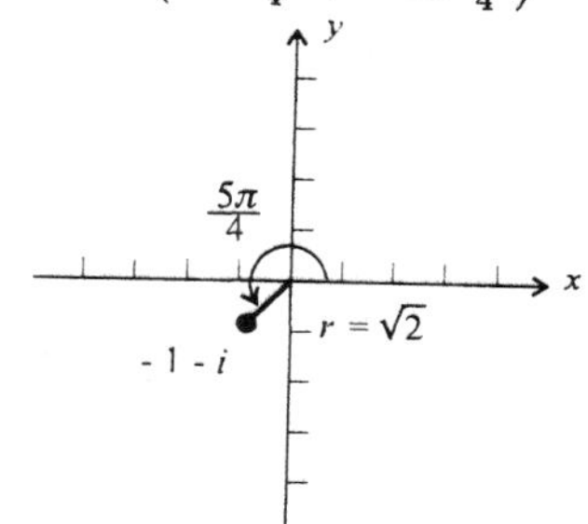

g. $-4 = 4\left(\cos\pi + i\sin\pi\right)$

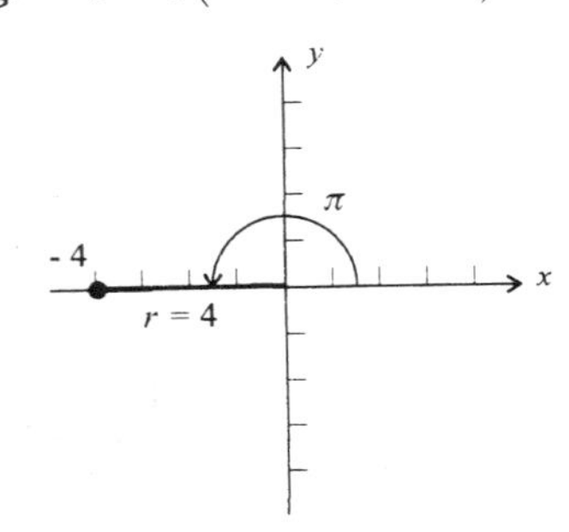

h. $-5i = 5\left(\cos\frac{3\pi}{2} + i\sin\frac{3\pi}{2}\right)$

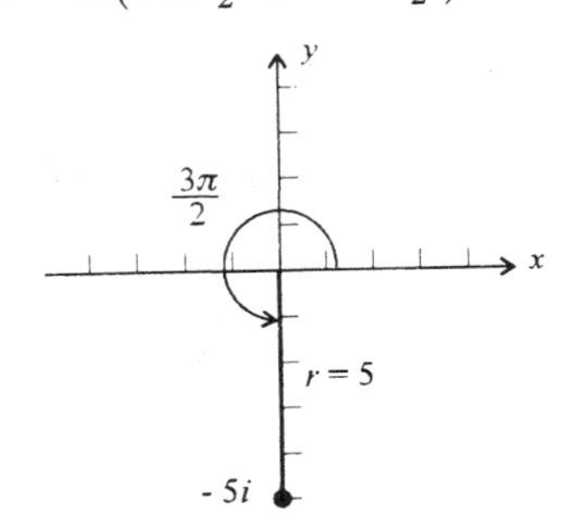

2. a. $4\left(\cos\frac{3\pi}{4} + i\sin\frac{3\pi}{4}\right) = -2\sqrt{2} + 2\sqrt{2}\,i$

   b. $3\left(\cos\frac{11\pi}{6} + i\sin\frac{11\pi}{6}\right) = \frac{3\sqrt{3}}{2} - \frac{3}{2}\,i$ c. $6\left(\cos\frac{2\pi}{3} + i\sin\frac{2\pi}{3}\right) = -3 + 3\sqrt{3}\,i$

   d. $30\left(\cos\frac{4\pi}{3} + i\sin\frac{4\pi}{3}\right) = -15 - 15\sqrt{3}\,i$

3. a. $-64\sqrt{3} - 64i$ b. $-i$ c. $512 + 512\sqrt{3}\,i$ d. -1 e. 1

   f. $256\sqrt{2} + 256\sqrt{2}\,i$ g. $-128 - 128\sqrt{3}\,i$ h. 1

6. a.

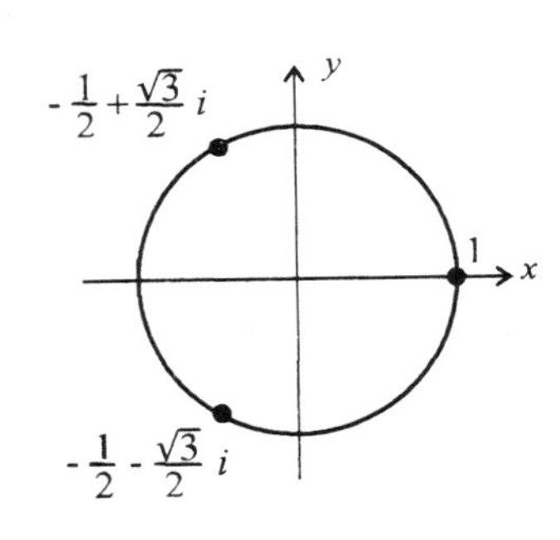

b.

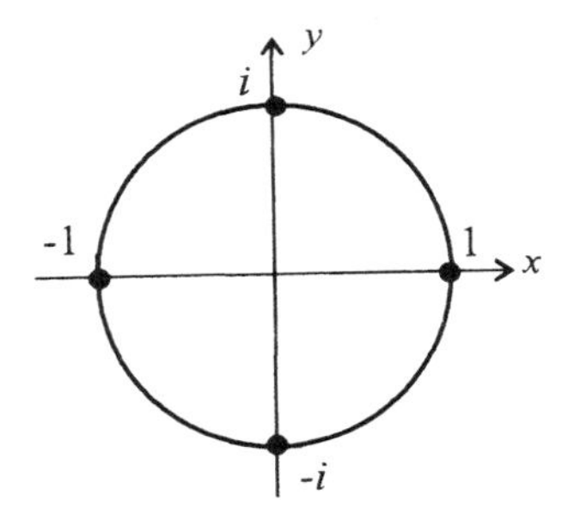

c.

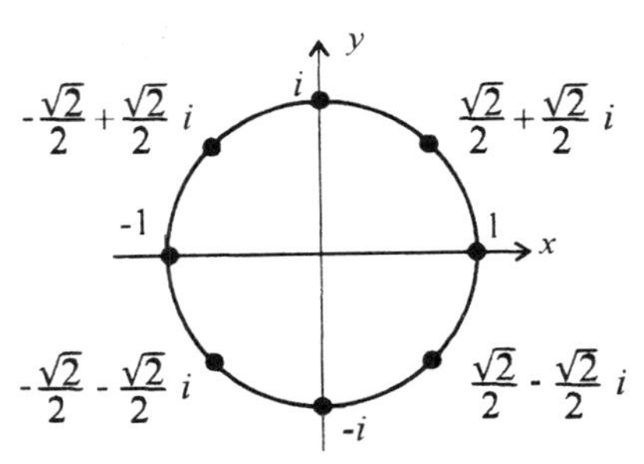

d.

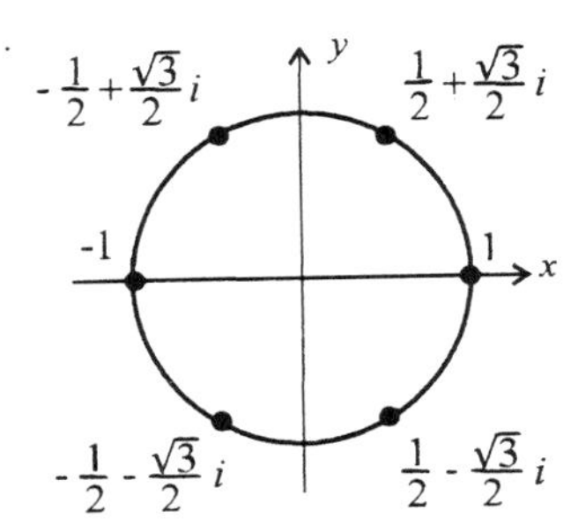

7. a. $\cos\frac{\pi}{18} + i\sin\frac{\pi}{18}$, $\cos\frac{13\pi}{18} + i\sin\frac{13\pi}{18}$, $\cos\frac{25\pi}{18} + i\sin\frac{25\pi}{18}$

   b. $\sqrt[6]{2}\left(\cos\frac{\pi}{4} + i\sin\frac{\pi}{4}\right)$, $\sqrt[6]{2}\left(\cos\frac{11\pi}{12} + i\sin\frac{11\pi}{12}\right)$, $\sqrt[6]{2}\left(\cos\frac{19\pi}{12} + i\sin\frac{19\pi}{12}\right)$

c. $\cos\frac{5\pi}{24}+i\sin\frac{5\pi}{24},\ \cos\frac{17\pi}{24}+i\sin\frac{17\pi}{24},\ \cos\frac{29\pi}{24}+i\sin\frac{29\pi}{24},\ \cos\frac{41\pi}{24}+i\sin\frac{41\pi}{24}$

d. $\cos\frac{5\pi}{12}+i\sin\frac{5\pi}{12},\ \cos\frac{11\pi}{12}+i\sin\frac{11\pi}{12},\ \cos\frac{17\pi}{12}+i\sin\frac{17\pi}{12},\ \cos\frac{23\pi}{12}+i\sin\frac{23\pi}{12}$

e. $2\left(\cos\frac{\pi}{4}+i\sin\frac{\pi}{4}\right),\ 2\left(\cos\frac{13\pi}{20}+i\sin\frac{13\pi}{20}\right),\ 2\left(\cos\frac{21\pi}{20}+i\sin\frac{21\pi}{20}\right),$
$2\left(\cos\frac{29\pi}{20}+i\sin\frac{29\pi}{20}\right),\ 2\left(\cos\frac{37\pi}{20}+i\sin\frac{37\pi}{20}\right)$

f. $2\left(\cos\frac{11\pi}{36}+i\sin\frac{11\pi}{36}\right),\ 2\left(\cos\frac{23\pi}{36}+i\sin\frac{23\pi}{36}\right),\ 2\left(\cos\frac{35\pi}{36}+i\sin\frac{35\pi}{36}\right),$
$2\left(\cos\frac{47\pi}{36}+i\sin\frac{47\pi}{36}\right),\ 2\left(\cos\frac{59\pi}{36}+i\sin\frac{59\pi}{36}\right),\ 2\left(\cos\frac{71\pi}{36}+i\sin\frac{71\pi}{36}\right)$

8. a. $\frac{3}{2}+\frac{3\sqrt{3}}{2}i,\ -3,\ \frac{3}{2}-\frac{3\sqrt{3}}{2}i$

b. $\sqrt{2},\ 1+i,\ \sqrt{2}i,\ -1+i,\ -\sqrt{2},\ -1-i,\ -\sqrt{2}i,\ 1-i$

c. $\frac{\sqrt{3}}{2}+\frac{1}{2}i,\ -\frac{\sqrt{3}}{2}+\frac{1}{2}i,\ -i$ d. $2i,\ -\sqrt{3}-i,\ \sqrt{3}-i$

e. $\frac{\sqrt{3}}{2}+\frac{1}{2}i,\ -\frac{1}{2}+\frac{\sqrt{3}}{2}i,\ -\frac{\sqrt{3}}{2}-\frac{1}{2}i,\ \frac{1}{2}-\frac{\sqrt{3}}{2}i$

f. $\frac{\sqrt[4]{18}}{2}+\frac{\sqrt[4]{2}}{2}i,\ -\frac{\sqrt[4]{2}}{2}+\frac{\sqrt[4]{18}}{2}i,\ -\frac{\sqrt[4]{18}}{2}-\frac{\sqrt[4]{2}}{2}i,\ \frac{\sqrt[4]{2}}{2}-\frac{\sqrt[4]{18}}{2}i$

g. $\frac{1}{2}+\frac{\sqrt{3}}{2}i,\ -\frac{\sqrt{3}}{2}+\frac{1}{2}i,\ -\frac{1}{2}-\frac{\sqrt{3}}{2}i,\ \frac{\sqrt{3}}{2}-\frac{1}{2}i$

h. $1+\sqrt{3}i,\ -\sqrt{3}+i,\ -1-\sqrt{3}i,\ \sqrt{3}-i$

11. $\left\langle\cos\frac{2\pi}{3}+i\sin\frac{2\pi}{3}\right\rangle=\left\{\cos\frac{2\pi}{3}+i\sin\frac{2\pi}{3},\ \cos\frac{4\pi}{3}+i\sin\frac{4\pi}{3},\ \cos 0+i\sin 0\right\}$

12. $\left\langle\cos\frac{3\pi}{2}+i\sin\frac{3\pi}{2}\right\rangle=\left\{\cos\frac{3\pi}{2}+i\sin\frac{3\pi}{2},\ \cos\pi+i\sin\pi,\ \cos\frac{\pi}{2}+i\sin\frac{\pi}{2},\ \cos 0+i\sin 0\right\}$

13. $\left\langle\cos\frac{5\pi}{3}+i\sin\frac{5\pi}{3}\right\rangle=\left\{\cos\frac{5\pi}{3}+i\sin\frac{5\pi}{3},\ \cos\frac{4\pi}{3}+i\sin\frac{4\pi}{3},\ \cos\pi+i\sin\pi,\right.$
$\left.\cos\frac{2\pi}{3}+i\sin\frac{2\pi}{3},\ \cos\frac{\pi}{3}+i\sin\frac{\pi}{3},\ \cos 0+i\sin 0\right\}$

14. $\left\langle\cos\frac{5\pi}{4}+i\sin\frac{5\pi}{4}\right\rangle=\left\{\cos\frac{5\pi}{4}+i\sin\frac{5\pi}{4},\cos\frac{\pi}{2}+i\sin\frac{\pi}{2},\cos\frac{7\pi}{4}+i\sin\frac{7\pi}{4},\cos\pi+i\sin\pi,\right.$

$\left.\cos\frac{\pi}{4}+i\sin\frac{\pi}{4},\cos\frac{3\pi}{2}+i\sin\frac{3\pi}{2},\cos\frac{3\pi}{4}+i\sin\frac{3\pi}{4},\cos 0+i\sin 0\right\}$

17. a. $\frac{1}{2}+\frac{\sqrt{3}}{2}i,\ \frac{1}{2}-\frac{\sqrt{3}}{2}i$

b. $\frac{\sqrt{2}}{2}+\frac{\sqrt{2}}{2}i,\ -\frac{\sqrt{2}}{2}+\frac{\sqrt{2}}{2}i,-\frac{\sqrt{2}}{2}-\frac{\sqrt{2}}{2}i,\ \frac{\sqrt{2}}{2}-\frac{\sqrt{2}}{2}i$

Exercises 8.1

1. a. $c_0x^0+c_1x^1+c_2x^2+c_3x^3$ or $c_0+c_1x+c_2x^2+c_3x^3$

b. $d_0x^0+d_1x^1+d_2x^2+d_3x^3+d_4x^4$ or $d_0+d_1x+d_2x^2+d_3x^3+d_4x^4$

c. $a_1x^1+a_2x^2+a_3x^3$ or $a_1x+a_2x^2+a_3x^3$ d. $x^2+x^3+x^4$

2. a. $\sum_{j=0}^{2} c_jx^j$ b. $\sum_{j=2}^{4} d_jx^j$ c. $\sum_{i=1}^{4} x^i$ d. $\sum_{k=3}^{5} x^k$

3. a. $2x^3+4x^2+3x+2$ b. $2x^2+4x+1$ c. $4x^2+2x$ d. $4x+2$

e. $2x^2+2x+3$ f. $2x^3+3x+6$ g. $4x^5+4x^2+7x+4$ h. $4x^5+x+6$

4. a. $2x^3+4x^2+2x+1$ b. x^2+4x c. $8x^5+8x^4+4x^3+8x^2+4x+6$
   d. $6x^4+6x^3+3x^2+3x$ e. $8x^5+8x^4+4x^3+5x^2+4x$
   f. $6x^4+8x^3+3x^2+x+4$ g. $2x^5+8x^4+7x^3+5x^2+7x$
   h. $3x^5+6x^4+6x+3$

5. a. The set S of all polynomials with zero constant term is nonempty, since it contains the zero polynomial. Both the sum and the product of polynomials with zero constant terms are again polynomials with zero constant terms, so S is closed under addition and multiplication. The additive inverse of a polynomial with zero constant term is also a polynomial with zero constant term, so S is a subring of $R[x]$.
   b. The subset is not a subring of $R[x]$ because it is not closed with respect to multiplication. The polynomials x^3 and x are in the subset, but the product $x^3 \cdot x = x^4$ is not in the subset.
   c. Let S be the set of all polynomials that have zero coefficients for all odd powers of x. Then x^2 is in S, so S is nonempty. For arbitrary

$$f(x)=\sum_{i=0}^{n} a_{2i}x^{2i} \quad \text{and} \quad g(x)=\sum_{i=0}^{m} b_{2i}x^{2i}$$

   in S, let k be the larger of n and m. Then

$$f(x)+g(x)=\sum_{i=0}^{k}(a_{2i}+b_{2i})x^{2i}$$

   has zero coefficients for all odd powers of x, and therefore is in S. Also,

$$f(x)g(x)=\sum_{i=0}^{m+n}\left(\sum_{p+q=i} a_{2p}b_{2q}\right)x^{2i}$$

   is in S, and

$$-f(x)=\sum_{i=0}^{n}(-a_{2i})x^{2i}$$

   is in S. Thus, S is a subring of $R[x]$.
   d. The subset is not a subring of $R[x]$ because it is not closed with respect to multiplication. The polynomials x^2 and x are in the subset, but $x^2 \cdot x = x^3$ is not in the subset.

6. a. The set S described in 5a is a subring of $R[x]$. Since a product of a polynomial with zero constant term and any other polynomial always has zero constant term, S is an ideal of $R[x]$.

b. The subset described in 5b is not an ideal of $R[x]$ since it is not a subring of $R[x]$.

c. The set S described in 5c is not an ideal of $R[x]$. The polynomial x^2 is in S, but the product $x(x^2) = x^3$ is not in S.

d. The subset described in 5d is not an ideal of $R[x]$ since it is not a subring of $R[x]$.

8. $x^2,\ x^2+1,\ x^2+2,\ x^2+x,\ x^2+x+1,\ x^2+x+2,\ x^2+2x,\ x^2+2x+1,$ $x^2+2x+2,\ 2x^2,\ 2x^2+1,\ 2x^2+2,\ 2x^2+x,\ 2x^2+x+1,\ 2x^2+x+2,$ $2x^2+2x,\ 2x^2+2x+1,\ 2x^2+2x+2$

9. a. $n^2(n-1)$ b. $n^m(n-1)$

10. b. $\ker\theta$ is the set of all polynomials in $R[x]$ that have zero constant term. (That is, $\ker\theta$ is the principal ideal generated by x in $R[x]$.)

14. $\ker\phi$ is the set of all polynomials in $\mathbf{Z}[x]$ that are multiples of k. (That is, $\ker\phi$ is the principal ideal generated by k in $\mathbf{Z}[x]$.)

16. $\ker\phi$ is the set of all polynomials $f(x) = a_0 + a_1x + \cdots + a_nx^n$ in $R[x]$ such that all coefficients a_i are in $\ker\theta$.

Exercises 8.2

1. $q(x) = 4x^2+3x+2,\ r(x) = 4$
2. $q(x) = 2x^2+2x+1,\ r(x) = 0$
3. $q(x) = x+2,\ r(x) = x^2+x$
4. $q(x) = 2x^2+x+1,\ r(x) = x^2+x$
5. $q(x) = 5x^2+3,\ r(x) = 2x+3$
6. $q(x) = 6x^2+3x+5,\ r(x) = 5x+2$
7. $d(x) = x+1$
8. $d(x) = x+1$
9. $d(x) = x+5$
10. $d(x) = 1$
11. $s(x) = x^2+2x+1,\ t(x) = x$
12. $s(x) = 2x^3+2x+2,\ t(x) = 2x+2$
13. $s(x) = x^2+2,\ t(x) = 4$
14. $s(x) = x^3+5x^2+6x+5,\ t(x) = x+1$

Exercises 8.3

1. a. $x^2-2 = (x-\sqrt{2})(x+\sqrt{2})$ is irreducible over $\mathbf{Q}$, reducible over $\mathbf{R}$ and $\mathbf{C}$.

b. $x^2+1 = (x+i)(x-i)$ is irreducible over $\mathbf{Q}$ and $\mathbf{R}$, reducible over $\mathbf{C}$.

c. $x^2+x-2 = (x+2)(x-1)$ is reducible over the fields $\mathbf{Q}, \mathbf{R}$, and $\mathbf{C}$.

d. $x^2+2x+2 = (x+1+i)(x+1-i)$ is irreducible over $\mathbf{Q}$ and $\mathbf{R}$, reducible over $\mathbf{C}$.

e. x^2+x+2 is irreducible over $\mathbf{Z}_3$ and $\mathbf{Z}_5$; $x^2+x+2 = (x+4)^2$ is reducible over $\mathbf{Z}_7$.

f. x^2+2x+2 is irreducible over $\mathbf{Z}_3$ and $\mathbf{Z}_7$; $x^2+2x+2=(x+4)(x+3)$ is reducible over $\mathbf{Z}_5$.

g. x^3-x^2+2x+2 is irreducible over $\mathbf{Z}_3$; $x^3-x^2+2x+2=(x+3)^3$ is reducible over $\mathbf{Z}_5$; $x^3-x^2+2x+2=(x+2)\left(x^2+4x+1\right)$ is reducible over $\mathbf{Z}_7$.

h. $x^4+2x^2+1=\left(x^2+1\right)^2$ is reducible over each of $\mathbf{Z}_3, \mathbf{Z}_5$, and $\mathbf{Z}_7$.

2. $x^2+1,\ x^2+x+2,\ x^2+2x+2$

3. a. $2(x+2)\left(x^2+3x+4\right)$ b. $3(x+1)\left(x^2+3x+4\right)$
c. $3(x+1)(x+2)(x+4)$ d. $2(x+1)(x+2)(x+4)$
e. $2(x+1)(x+2)\left(x^2+3\right)$ f. $3(x+2)(x+4)\left(x^2+2\right)$
g. $(x+3)^2\left(x^2+2\right)$ h. $(x+3)^2\left(x^2+3\right)$

7. $x^4+5x^2+4=\left(x^2+1\right)\left(x^2+4\right)$ is reducible over $\mathbf{R}$ and has no zeros in the field of real numbers.

Exercises 8.4

1. a. $f(x)=x^2-(3+2i)x+6i,\ g(x)=x^3-3x^2+4x-12$
b. $f(x)=x^2-(4-3i)x-12i,\ g(x)=x^3-4x^2+9x-36$
c. $f(x)=x^2-(3-i)x+(2-2i),\ g(x)=x^3-4x^2+6x-4$
d. $f(x)=x^2-(5-i)x+(6-3i),\ g(x)=x^3-7x^2+17x-15$
e. $f(x)=x^2-(1+5i)x-(6-3i),\ g(x)=x^4-2x^3+14x^2-18x+45$
f. $f(x)=x^2-2x+(1+2i),\ g(x)=x^4-4x^3+6x^2-4x+5$
g. $f(x)=x^3-3x^2+(3-2i)x-(1-2i),\ g(x)=x^5-5x^4+10x^3-10x^2+9x-5$
h. $f(x)=x^3-5x^2+(7+3i)x-(2+6i),\ g(x)=x^5-8x^4+23x^3-28x^2+22x-20$

2. a. $1+i, 2$ b. $1+i, -3$ c. $i, \frac{-1+i\sqrt{3}}{2}, \frac{-1-i\sqrt{3}}{2}$ d. $-2i, -1, -2$

3. $\frac{5}{2}, -1$ **4.** $-2, -4, -\frac{1}{3}$ **5.** $\frac{3}{2}, -1$ **6.** $2, -\frac{5}{2}$ **7.** $-2, \frac{1+i\sqrt{3}}{2}, \frac{1-i\sqrt{3}}{2}$

8. $\frac{1}{3}, 1+i, 1-i$ **9.** $1, \frac{1}{3}, -2$ **10.** $2, -1, -\frac{1}{3}$ **11.** $-1, \frac{1}{2}, -\frac{4}{3}$ **12.** $\frac{2}{3}, -2, -\frac{5}{3}$

13. $(x-1)(x+2)\left(x^2-2x+2\right)$ **14.** $2(x+1)\left(x-\frac{3}{2}\right)\left(x^2-5\right)$

15. $2\left(x-\frac{1}{2}\right)(x+3)\left(x^2-2\right)$ **16.** $6\left(x+\frac{1}{2}\right)\left(x-\frac{4}{3}\right)\left(x^2+x+2\right)$

17. a. Let $f(x)=3+9x+x^3$. The prime integer 3 divides all the coefficients of $f(x)$ except the leading coefficient $a_n=1$, and 3^2 does not divide $a_0=3$. Thus $f(x)$ is irreducible, by Eisenstein's Criterion.

b. Let $f(x) = 7 - 14x + 28x^2 + x^3$. The prime integer 7 divides all the coefficients of $f(x)$ except the leading coefficient $a_n = 1$, and 7^2 does not divide $a_0 = 7$. Thus $f(x)$ is irreducible, by Eisenstein's Criterion.

c. Let $f(x) = 3 - 27x^2 + 2x^5$. The prime integer 3 divides all the coefficients of $f(x)$ except the leading coefficient $a_n = 2$, and 3^2 does not divide $a_0 = 3$. Thus $f(x)$ is irreducible, by Eisenstein's Criterion.

d. Let $f(x) = 6 + 12x^2 - 27x^3 + 10x^5$. The prime integer 3 divides all the coefficients of $f(x)$ except the leading coefficient $a_n = 10$, and 3^2 does not divide $a_0 = 6$. Thus $f(x)$ is irreducible, by Eisenstein's Criterion.

Exercises 8.5

1. $\sqrt[3]{25} + \sqrt[3]{5}, \quad -\dfrac{\sqrt[3]{25} + \sqrt[3]{5}}{2} \pm i\sqrt{3}\,\dfrac{\sqrt[3]{25} - \sqrt[3]{5}}{2}$

2. $-\sqrt[3]{3} - \sqrt[3]{9}, \quad \dfrac{\sqrt[3]{3} + \sqrt[3]{9}}{2} \pm i\sqrt{3}\,\dfrac{\sqrt[3]{3} - \sqrt[3]{9}}{2}$

3. $\sqrt[3]{16} + \sqrt[3]{4}, \quad -\dfrac{\sqrt[3]{16} + \sqrt[3]{4}}{2} \pm i\sqrt{3}\,\dfrac{\sqrt[3]{16} - \sqrt[3]{4}}{2}$

4. $\sqrt[3]{25} - \sqrt[3]{5}, \quad \dfrac{\sqrt[3]{5} - \sqrt[3]{25}}{2} \pm i\sqrt{3}\,\dfrac{\sqrt[3]{5} + \sqrt[3]{25}}{2}$

5. $\sqrt[3]{4} + \sqrt[3]{2}, \quad -\dfrac{\sqrt[3]{4} + \sqrt[3]{2}}{2} \pm i\sqrt{3}\,\dfrac{\sqrt[3]{4} - \sqrt[3]{2}}{2}$

6. $\sqrt[3]{4} - \sqrt[3]{2}, \quad \dfrac{\sqrt[3]{2} - \sqrt[3]{4}}{2} \pm i\sqrt{3}\,\dfrac{\sqrt[3]{2} + \sqrt[3]{4}}{2}$

7. $\sqrt[3]{3} - \sqrt[3]{9}, \quad \dfrac{\sqrt[3]{9} - \sqrt[3]{3}}{2} \pm i\sqrt{3}\,\dfrac{\sqrt[3]{9} + \sqrt[3]{3}}{2}$

8. $\sqrt[3]{9} - \sqrt[3]{3}, \quad \dfrac{\sqrt[3]{3} - \sqrt[3]{9}}{2} \pm i\sqrt{3}\,\dfrac{\sqrt[3]{3} + \sqrt[3]{9}}{2}$

9. $\dfrac{2\sqrt[3]{2} - \sqrt[3]{4}}{2}, \quad \dfrac{\sqrt[3]{4} - 2\sqrt[3]{2}}{4} \pm i\sqrt{3}\,\dfrac{\sqrt[3]{4} + 2\sqrt[3]{2}}{4}$

10. $\dfrac{2\sqrt[3]{2} + \sqrt[3]{4}}{2}, \quad -\dfrac{2\sqrt[3]{2} + \sqrt[3]{4}}{4} \pm i\sqrt{3}\,\dfrac{2\sqrt[3]{2} - \sqrt[3]{4}}{4}$

11. $\sqrt[3]{49} - \sqrt[3]{7} + 2, \quad \dfrac{\sqrt[3]{7} - \sqrt[3]{49} + 4}{2} \pm i\sqrt{3}\,\dfrac{\sqrt[3]{7} + \sqrt[3]{49}}{2}$

12. $\sqrt[3]{18} - \sqrt[3]{12} - 1, \quad \dfrac{\sqrt[3]{12} - \sqrt[3]{18} - 2}{2} \pm i\sqrt{3}\,\dfrac{\sqrt[3]{12} + \sqrt[3]{18}}{2}$

13. $\sqrt[3]{18} - \sqrt[3]{12} - \frac{1}{2}, \quad \frac{\sqrt[3]{12} - \sqrt[3]{18} - 1}{2} \pm i\sqrt{3}\,\frac{\sqrt[3]{12} + \sqrt[3]{18}}{2}$

14. $\frac{2\sqrt[3]{2} - 2\sqrt[3]{4} + 1}{2}, \quad \frac{\sqrt[3]{4} - \sqrt[3]{2} + 1}{2} \pm i\sqrt{3}\,\frac{\sqrt[3]{4} + \sqrt[3]{2}}{2}$ **15**. $1 \pm i, \quad -1 \pm i\sqrt{2}$

16. $-1 \pm \sqrt{2}, \quad 1 \pm i\sqrt{2}$ **17**. $\pm i, \quad -2 \pm \sqrt{2}$ **18**. $2 \pm \sqrt{2}, \quad \pm i\sqrt{2}$

Exercises 8.6

1. a. Let $P = (p(x))$ and $\alpha = x + P$ in $\mathbf{Z}_3[x]/P$. The elements of $\mathbf{Z}_3[x]/P$ are

$$\{0,\ 1,\ 2,\ \alpha,\ \alpha+1,\ \alpha+2,\ 2\alpha,\ 2\alpha+1,\ 2\alpha+2\}$$

where $0 = 0 + P$, $1 = 1 + P$, and $2 = 2 + P$. Addition and multiplication tables are as follows.

$+$	0	1	2	α	$\alpha+1$	$\alpha+2$	2α	$2\alpha+1$	$2\alpha+2$
0	0	1	2	α	$\alpha+1$	$\alpha+2$	2α	$2\alpha+1$	$2\alpha+2$
1	1	2	0	$\alpha+1$	$\alpha+2$	α	$2\alpha+1$	$2\alpha+2$	2α
2	2	0	1	$\alpha+2$	α	$\alpha+1$	$2\alpha+2$	2α	$2\alpha+1$
α	α	$\alpha+1$	$\alpha+2$	2α	$2\alpha+1$	$2\alpha+2$	0	1	2
$\alpha+1$	$\alpha+1$	$\alpha+2$	α	$2\alpha+1$	$2\alpha+2$	2α	1	2	0
$\alpha+2$	$\alpha+2$	α	$\alpha+1$	$2\alpha+2$	2α	$2\alpha+1$	2	0	1
2α	2α	$2\alpha+1$	$2\alpha+2$	0	1	2	α	$\alpha+1$	$\alpha+2$
$2\alpha+1$	$2\alpha+1$	$2\alpha+2$	2α	1	2	0	$\alpha+1$	$\alpha+2$	α
$2\alpha+2$	$2\alpha+2$	2α	$2\alpha+1$	2	0	1	$\alpha+2$	α	$\alpha+1$

$\cdot$	0	1	2	α	$\alpha+1$	$\alpha+2$	2α	$2\alpha+1$	$2\alpha+2$
0	0	0	0	0	0	0	0	0	0
1	0	1	2	α	$\alpha+1$	$\alpha+2$	2α	$2\alpha+1$	$2\alpha+2$
2	0	2	1	2α	$2\alpha+2$	$2\alpha+1$	α	$\alpha+2$	$\alpha+1$
α	0	α	2α	$2\alpha+1$	1	$\alpha+1$	$\alpha+2$	$2\alpha+2$	2
$\alpha+1$	0	$\alpha+1$	$2\alpha+2$	1	$\alpha+2$	2α	2	α	$2\alpha+1$
$\alpha+2$	0	$\alpha+2$	$2\alpha+1$	$\alpha+1$	2α	2	$2\alpha+2$	1	α
2α	0	2α	α	$\alpha+2$	2	$2\alpha+2$	$2\alpha+1$	$\alpha+1$	1
$2\alpha+1$	0	$2\alpha+1$	$\alpha+2$	$2\alpha+2$	α	1	$\alpha+1$	2	2α
$2\alpha+2$	0	$2\alpha+2$	$\alpha+1$	2	$2\alpha+1$	α	1	2α	$\alpha+2$

1. b. Let $P = (p(x))$ and $\alpha = x + P$ in $\mathbf{Z}_3[x]/P$. The elements of $\mathbf{Z}_3[x]/P$ are

$$\{0,\ 1,\ 2,\ \alpha,\ \alpha+1,\ \alpha+2,\ 2\alpha,\ 2\alpha+1,\ 2\alpha+2\}$$

where $0 = 0 + P$, $1 = 1 + P$, and $2 = 2 + P$. Addition and multiplication tables are as follows.

$+$	0	1	2	α	$\alpha+1$	$\alpha+2$	2α	$2\alpha+1$	$2\alpha+2$
0	0	1	2	α	$\alpha+1$	$\alpha+2$	2α	$2\alpha+1$	$2\alpha+2$
1	1	2	0	$\alpha+1$	$\alpha+2$	α	$2\alpha+1$	$2\alpha+2$	2α
2	2	0	1	$\alpha+2$	α	$\alpha+1$	$2\alpha+2$	2α	$2\alpha+1$
α	α	$\alpha+1$	$\alpha+2$	2α	$2\alpha+1$	$2\alpha+2$	0	1	2
$\alpha+1$	$\alpha+1$	$\alpha+2$	α	$2\alpha+1$	$2\alpha+2$	2α	1	2	0
$\alpha+2$	$\alpha+2$	α	$\alpha+1$	$2\alpha+2$	2α	$2\alpha+1$	2	0	1
2α	2α	$2\alpha+1$	$2\alpha+2$	0	1	2	α	$\alpha+1$	$\alpha+2$
$2\alpha+1$	$2\alpha+1$	$2\alpha+2$	2α	1	2	0	$\alpha+1$	$\alpha+2$	α
$2\alpha+2$	$2\alpha+2$	2α	$2\alpha+1$	2	0	1	$\alpha+2$	α	$\alpha+1$

$\cdot$	0	1	2	α	$\alpha+1$	$\alpha+2$	2α	$2\alpha+1$	$2\alpha+2$
0	0	0	0	0	0	0	0	0	0
1	0	1	2	α	$\alpha+1$	$\alpha+2$	2α	$2\alpha+1$	$2\alpha+2$
2	0	2	1	2α	$2\alpha+2$	$2\alpha+1$	α	$\alpha+2$	$\alpha+1$
α	0	α	2α	2	$\alpha+2$	$2\alpha+2$	1	$\alpha+1$	$2\alpha+1$
$\alpha+1$	0	$\alpha+1$	$2\alpha+2$	$\alpha+2$	2α	1	$2\alpha+1$	2	α
$\alpha+2$	0	$\alpha+2$	$2\alpha+1$	$2\alpha+2$	1	α	$\alpha+1$	2α	2
2α	0	2α	α	1	$2\alpha+1$	$\alpha+1$	2	$2\alpha+2$	$\alpha+2$
$2\alpha+1$	0	$2\alpha+1$	$\alpha+2$	$\alpha+1$	2	2α	$2\alpha+2$	α	1
$2\alpha+2$	0	$2\alpha+2$	$\alpha+1$	$2\alpha+1$	α	2	$\alpha+2$	1	2α

2. **a.** $\mathbf{Z}_2[x]/(p(x)) = \{0,\ 1,\ \alpha,\ \alpha+1\}$ is a field.

$+$	0	1	α	$\alpha+1$
0	0	1	α	$\alpha+1$
1	1	0	$\alpha+1$	α
α	α	$\alpha+1$	0	1
$\alpha+1$	$\alpha+1$	α	1	0

$\cdot$	0	1	α	$\alpha+1$
0	0	0	0	0
1	0	1	α	$\alpha+1$
α	0	α	$\alpha+1$	1
$\alpha+1$	0	$\alpha+1$	1	α

2. **b.** $\mathbf{Z}_2[x]/(p(x)) = \{0,\ 1,\ \alpha,\ \alpha+1,\ \alpha^2,\ \alpha^2+1,\ \alpha^2+\alpha,\ \alpha^2+\alpha+1\}$ is not a field since $\alpha+1$ does not have a multiplicative inverse.

$+$	0	1	α	$\alpha+1$	α^2	α^2+1	$\alpha^2+\alpha$	$\alpha^2+\alpha+1$
0	0	1	α	$\alpha+1$	α^2	α^2+1	$\alpha^2+\alpha$	$\alpha^2+\alpha+1$
1	1	0	$\alpha+1$	α	α^2+1	α^2	$\alpha^2+\alpha+1$	$\alpha^2+\alpha$
α	α	$\alpha+1$	0	1	$\alpha^2+\alpha$	$\alpha^2+\alpha+1$	α^2	α^2+1
$\alpha+1$	$\alpha+1$	α	1	0	$\alpha^2+\alpha+1$	$\alpha^2+\alpha$	α^2+1	α^2
α^2	α^2	α^2+1	$\alpha^2+\alpha$	$\alpha^2+\alpha+1$	0	1	α	$\alpha+1$
α^2+1	α^2+1	α^2	$\alpha^2+\alpha+1$	$\alpha^2+\alpha$	1	0	$\alpha+1$	α
$\alpha^2+\alpha$	$\alpha^2+\alpha$	$\alpha^2+\alpha+1$	α^2	α^2+1	α	$\alpha+1$	0	1
$\alpha^2+\alpha+1$	$\alpha^2+\alpha+1$	$\alpha^2+\alpha$	α^2+1	α^2	$\alpha+1$	α	1	0

$\cdot$	0	1	α	$\alpha+1$	α^2	α^2+1	$\alpha^2+\alpha$	$\alpha^2+\alpha+1$
0	0	0	0	0	0	0	0	0
1	0	1	α	$\alpha+1$	α^2	α^2+1	$\alpha^2+\alpha$	$\alpha^2+\alpha+1$
α	0	α	α^2	$\alpha^2+\alpha$	1	$\alpha+1$	α^2+1	$\alpha^2+\alpha+1$
$\alpha+1$	0	$\alpha+1$	$\alpha^2+\alpha$	α^2+1	α^2+1	$\alpha^2+\alpha$	$\alpha+1$	0
α^2	0	α^2	1	α^2+1	α	$\alpha^2+\alpha$	$\alpha+1$	$\alpha^2+\alpha+1$
α^2+1	0	α^2+1	$\alpha+1$	$\alpha^2+\alpha$	$\alpha^2+\alpha$	$\alpha+1$	α^2+1	0
$\alpha^2+\alpha$	0	$\alpha^2+\alpha$	α^2+1	$\alpha+1$	$\alpha+1$	α^2+1	$\alpha^2+\alpha$	0
$\alpha^2+\alpha+1$	0	$\alpha^2+\alpha+1$	$\alpha^2+\alpha+1$	0	$\alpha^2+\alpha+1$	0	0	$\alpha^2+\alpha+1$

2. c. $\mathbf{Z}_2[x]/(p(x)) = \{0,\ 1,\ \alpha,\ \alpha+1,\ \alpha^2,\ \alpha^2+1,\ \alpha^2+\alpha,\ \alpha^2+\alpha+1\}$ is a field.

$+$	0	1	α	$\alpha+1$	α^2	α^2+1	$\alpha^2+\alpha$	$\alpha^2+\alpha+1$
0	0	1	α	$\alpha+1$	α^2	α^2+1	$\alpha^2+\alpha$	$\alpha^2+\alpha+1$
1	1	0	$\alpha+1$	α	α^2+1	α^2	$\alpha^2+\alpha+1$	$\alpha^2+\alpha$
α	α	$\alpha+1$	0	1	$\alpha^2+\alpha$	$\alpha^2+\alpha+1$	α^2	α^2+1
$\alpha+1$	$\alpha+1$	α	1	0	$\alpha^2+\alpha+1$	$\alpha^2+\alpha$	α^2+1	α^2
α^2	α^2	α^2+1	$\alpha^2+\alpha$	$\alpha^2+\alpha+1$	0	1	α	$\alpha+1$
α^2+1	α^2+1	α^2	$\alpha^2+\alpha+1$	$\alpha^2+\alpha$	1	0	$\alpha+1$	α
$\alpha^2+\alpha$	$\alpha^2+\alpha$	$\alpha^2+\alpha+1$	α^2	α^2+1	α	$\alpha+1$	0	1
$\alpha^2+\alpha+1$	$\alpha^2+\alpha+1$	$\alpha^2+\alpha$	α^2+1	α^2	$\alpha+1$	α	1	0

$\cdot$	0	1	α	$\alpha+1$	α^2	α^2+1	$\alpha^2+\alpha$	$\alpha^2+\alpha+1$
0	0	0	0	0	0	0	0	0
1	0	1	α	$\alpha+1$	α^2	α^2+1	$\alpha^2+\alpha$	$\alpha^2+\alpha+1$
α	0	α	α^2	$\alpha^2+\alpha$	$\alpha+1$	1	$\alpha^2+\alpha+1$	α^2+1
$\alpha+1$	0	$\alpha+1$	$\alpha^2+\alpha$	α^2+1	$\alpha^2+\alpha+1$	α^2	1	α
α^2	0	α^2	$\alpha+1$	$\alpha^2+\alpha+1$	$\alpha^2+\alpha$	α	α^2+1	1
α^2+1	0	α^2+1	1	α^2	α	$\alpha^2+\alpha+1$	$\alpha+1$	$\alpha^2+\alpha$
$\alpha^2+\alpha$	0	$\alpha^2+\alpha$	$\alpha^2+\alpha+1$	1	α^2+1	$\alpha+1$	α	α^2
$\alpha^2+\alpha+1$	0	$\alpha^2+\alpha+1$	α^2+1	α	1	$\alpha^2+\alpha$	α^2	$\alpha+1$

2. d. $\mathbf{Z}_2[x]/(p(x)) = \{0,\ 1,\ \alpha,\ \alpha+1,\ \alpha^2,\ \alpha^2+1,\ \alpha^2+\alpha,\ \alpha^2+\alpha+1\}$ is a field.

$+$	0	1	α	$\alpha+1$	α^2	α^2+1	$\alpha^2+\alpha$	$\alpha^2+\alpha+1$
0	0	1	α	$\alpha+1$	α^2	α^2+1	$\alpha^2+\alpha$	$\alpha^2+\alpha+1$
1	1	0	$\alpha+1$	α	α^2+1	α^2	$\alpha^2+\alpha+1$	$\alpha^2+\alpha$
α	α	$\alpha+1$	0	1	$\alpha^2+\alpha$	$\alpha^2+\alpha+1$	α^2	α^2+1
$\alpha+1$	$\alpha+1$	α	1	0	$\alpha^2+\alpha+1$	$\alpha^2+\alpha$	α^2+1	α^2
α^2	α^2	α^2+1	$\alpha^2+\alpha$	$\alpha^2+\alpha+1$	0	1	α	$\alpha+1$
α^2+1	α^2+1	α^2	$\alpha^2+\alpha+1$	$\alpha^2+\alpha$	1	0	$\alpha+1$	α
$\alpha^2+\alpha$	$\alpha^2+\alpha$	$\alpha^2+\alpha+1$	α^2	α^2+1	α	$\alpha+1$	0	1
$\alpha^2+\alpha+1$	$\alpha^2+\alpha+1$	$\alpha^2+\alpha$	α^2+1	α^2	$\alpha+1$	α	1	0

$\cdot$	0	1	α	$\alpha+1$	α^2	α^2+1	$\alpha^2+\alpha$	$\alpha^2+\alpha+1$
0	0	0	0	0	0	0	0	0
1	0	1	α	$\alpha+1$	α^2	α^2+1	$\alpha^2+\alpha$	$\alpha^2+\alpha+1$
α	0	α	α^2	$\alpha^2+\alpha$	α^2+1	$\alpha^2+\alpha+1$	1	$\alpha+1$
$\alpha+1$	0	$\alpha+1$	$\alpha^2+\alpha$	α^2+1	1	α	$\alpha^2+\alpha+1$	α^2
α^2	0	α^2	α^2+1	1	$\alpha^2+\alpha+1$	$\alpha+1$	α	$\alpha^2+\alpha$
α^2+1	0	α^2+1	$\alpha^2+\alpha+1$	α	$\alpha+1$	$\alpha^2+\alpha$	α^2	1
$\alpha^2+\alpha$	0	$\alpha^2+\alpha$	1	$\alpha^2+\alpha+1$	α	α^2	$\alpha+1$	α^2+1
$\alpha^2+\alpha+1$	0	$\alpha^2+\alpha+1$	$\alpha+1$	α^2	$\alpha^2+\alpha$	1	α^2+1	α

2. e. $\mathbf{Z}_3[x]/(p(x)) = \{0, 1, 2, \alpha, \alpha+1, \alpha+2, 2\alpha, 2\alpha+1, 2\alpha+2\}$ is not a field since $\alpha+2$ does not have a multiplicative inverse.

$+$	0	1	2	α	$\alpha+1$	$\alpha+2$	2α	$2\alpha+1$	$2\alpha+2$
0	0	1	2	α	$\alpha+1$	$\alpha+2$	2α	$2\alpha+1$	$2\alpha+2$
1	1	2	0	$\alpha+1$	$\alpha+2$	α	$2\alpha+1$	$2\alpha+2$	2α
2	2	0	1	$\alpha+2$	α	$\alpha+1$	$2\alpha+2$	2α	$2\alpha+1$
α	α	$\alpha+1$	$\alpha+2$	2α	$2\alpha+1$	$2\alpha+2$	0	1	2
$\alpha+1$	$\alpha+1$	$\alpha+2$	α	$2\alpha+1$	$2\alpha+2$	2α	1	2	0
$\alpha+2$	$\alpha+2$	α	$\alpha+1$	$2\alpha+2$	2α	$2\alpha+1$	2	0	1
2α	2α	$2\alpha+1$	$2\alpha+2$	0	1	2	α	$\alpha+1$	$\alpha+2$
$2\alpha+1$	$2\alpha+1$	$2\alpha+2$	2α	1	2	0	$\alpha+1$	$\alpha+2$	α
$2\alpha+2$	$2\alpha+2$	2α	$2\alpha+1$	2	0	1	$\alpha+2$	α	$\alpha+1$

$\cdot$	0	1	2	α	$\alpha+1$	$\alpha+2$	2α	$2\alpha+1$	$2\alpha+2$
0	0	0	0	0	0	0	0	0	0
1	0	1	2	α	$\alpha+1$	$\alpha+2$	2α	$2\alpha+1$	$2\alpha+2$
2	0	2	1	2α	$2\alpha+2$	$2\alpha+1$	α	$\alpha+2$	$\alpha+1$
α	0	α	2α	$2\alpha+2$	2	$\alpha+2$	$\alpha+1$	$2\alpha+1$	1
$\alpha+1$	0	$\alpha+1$	$2\alpha+2$	2	α	$2\alpha+1$	1	$\alpha+2$	2α
$\alpha+2$	0	$\alpha+2$	$2\alpha+1$	$\alpha+2$	$2\alpha+1$	0	$2\alpha+1$	0	$\alpha+2$
2α	0	2α	α	$\alpha+1$	1	$2\alpha+1$	$2\alpha+2$	$\alpha+2$	2
$2\alpha+1$	0	$2\alpha+1$	$\alpha+2$	$2\alpha+1$	$\alpha+2$	0	$\alpha+2$	0	$2\alpha+1$
$2\alpha+2$	0	$2\alpha+2$	$\alpha+1$	1	2α	$\alpha+2$	2	$2\alpha+1$	α

2. f. $\mathbf{Z}_3[x]/(p(x)) = \{0, 1, 2, \alpha, \alpha+1, \alpha+2, 2\alpha, 2\alpha+1, 2\alpha+2\}$ is not a field since $\alpha+1$ does not have a multiplicative inverse.

$+$	0	1	2	α	$\alpha+1$	$\alpha+2$	2α	$2\alpha+1$	$2\alpha+2$
0	0	1	2	α	$\alpha+1$	$\alpha+2$	2α	$2\alpha+1$	$2\alpha+2$
1	1	2	0	$\alpha+1$	$\alpha+2$	α	$2\alpha+1$	$2\alpha+2$	2α
2	2	0	1	$\alpha+2$	α	$\alpha+1$	$2\alpha+2$	2α	$2\alpha+1$
α	α	$\alpha+1$	$\alpha+2$	2α	$2\alpha+1$	$2\alpha+2$	0	1	2
$\alpha+1$	$\alpha+1$	$\alpha+2$	α	$2\alpha+1$	$2\alpha+2$	2α	1	2	0
$\alpha+2$	$\alpha+2$	α	$\alpha+1$	$2\alpha+2$	2α	$2\alpha+1$	2	0	1
2α	2α	$2\alpha+1$	$2\alpha+2$	0	1	2	α	$\alpha+1$	$\alpha+2$
$2\alpha+1$	$2\alpha+1$	$2\alpha+2$	2α	1	2	0	$\alpha+1$	$\alpha+2$	α
$2\alpha+2$	$2\alpha+2$	2α	$2\alpha+1$	2	0	1	$\alpha+2$	α	$\alpha+1$

$\cdot$	0	1	2	α	$\alpha+1$	$\alpha+2$	2α	$2\alpha+1$	$2\alpha+2$
0	0	0	0	0	0	0	0	0	0
1	0	1	2	α	$\alpha+1$	$\alpha+2$	2α	$2\alpha+1$	$2\alpha+2$
2	0	2	1	2α	$2\alpha+2$	$2\alpha+1$	α	$\alpha+2$	$\alpha+1$
α	0	α	2α	1	$\alpha+1$	$2\alpha+1$	2	$\alpha+2$	$2\alpha+2$
$\alpha+1$	0	$\alpha+1$	$2\alpha+2$	$\alpha+1$	$2\alpha+2$	0	$2\alpha+2$	0	$\alpha+1$
$\alpha+2$	0	$\alpha+2$	$2\alpha+1$	$2\alpha+1$	0	$\alpha+2$	$\alpha+2$	$2\alpha+1$	0
2α	0	2α	α	2	$2\alpha+2$	$\alpha+2$	1	$2\alpha+1$	$\alpha+1$
$2\alpha+1$	0	$2\alpha+1$	$\alpha+2$	$\alpha+2$	0	$2\alpha+1$	$2\alpha+1$	$\alpha+2$	0
$2\alpha+2$	0	$2\alpha+2$	$\alpha+1$	$2\alpha+2$	$\alpha+1$	0	$\alpha+1$	0	$2\alpha+2$

3. a. We have $p(0) = 1, p(1) = 1$, and $p(2) = 2$. Therefore $p(x)$ is irreducible, by Theorem 8.20.

b. $$\begin{aligned}(a_0 + a_1\alpha + a_2\alpha^2)(b_0 + b_1\alpha + b_2\alpha^2) &= (a_0b_0 + 2a_1b_2 + 2a_2b_1 + 2a_2b_2) + (a_0b_1 + a_1b_0 + 2a_2b_2)\alpha \\ &\quad + (a_0b_2 + a_1b_1 + a_1b_2 + a_2b_0 + a_2b_1 + a_2b_2)\alpha^2\end{aligned}$$

c. $(\alpha^2 + \alpha + 2)^{-1} = \alpha + 1$

4. a. We have $p(0) = 1, p(1) = 2$, and $p(2) = 2$. Therefore $p(x)$ is irreducible, by Theorem 8.20.

   b. $\left(a_0 + a_1\alpha + a_2\alpha^2\right)\left(b_0 + b_1\alpha + b_2\alpha^2\right)$

$$= (a_0b_0 + 2a_1b_2 + 2a_2b_1 + a_2b_2) + (a_0b_1 + a_1b_0 + a_1b_2 + a_2b_1 + a_2b_2)\,\alpha + (a_0b_2 + a_1b_1 + a_2b_0 + 2a_1b_2 + 2a_2b_1 + 2a_2b_2)\,\alpha^2$$

   c. $\left(\alpha^2 + 2\alpha + 1\right)^{-1} = \alpha + 2$

5. a. We have $p(0) = 1, p(1) = 3$, $p(2) = 1, p(3) = 1$ and $p(4) = 4$. Therefore $p(x)$ is irreducible, by Theorem 8.20.

   b. $\left(a_0 + a_1\alpha + a_2\alpha^2\right)\left(b_0 + b_1\alpha + b_2\alpha^2\right)$

$$= (a_0b_0 + 4a_1b_2 + 4a_2b_1) + (a_0b_1 + a_1b_0 + 4a_1b_2 + 4a_2b_1 + 4a_2b_2)\,\alpha + (a_0b_2 + a_1b_1 + a_2b_0 + 4a_2b_2)\,\alpha^2$$

   c. $\left(\alpha^2 + 4\alpha\right)^{-1} = 4\alpha^2 + 3\alpha + 2$

6. a. We have $p(0) = 1, p(1) = 3, p(2) = 3, p(3) = 2$ and $p(4) = 1$. Therefore $p(x)$ is irreducible, by Theorem 8.20.

   b. $\left(a_0 + a_1\alpha + a_2\alpha^2\right)\left(b_0 + b_1\alpha + b_2\alpha^2\right)$

$$= (a_0b_0 + 4a_1b_2 + 4a_2b_1 + a_2b_2) + (a_0b_1 + a_1b_0 + 4a_2b_2)\,\alpha + (a_0b_2 + a_1b_1 + a_2b_0 + 4a_1b_2 + 4a_2b_1 + a_2b_2)\,\alpha^2$$

   c. $\left(\alpha^2 + 2\alpha + 3\right)^{-1} = 2\alpha^2 + 1$

7. a. $0, 1, 2, \alpha, \alpha+1, \alpha+2, 2\alpha, 2\alpha+1, 2\alpha+2, \alpha^2, \alpha^2+1, \alpha^2+2, 2\alpha^2, 2\alpha^2+1, 2\alpha^2+2, \alpha^2+\alpha, \alpha^2+\alpha+1, \alpha^2+\alpha+2, 2\alpha^2+\alpha, 2\alpha^2+\alpha+1, 2\alpha^2+\alpha+2, \alpha^2+2\alpha, \alpha^2+2\alpha+1, \alpha^2+2\alpha+2, 2\alpha^2+2\alpha, 2\alpha^2+2\alpha+1, 2\alpha^2+2\alpha+2$

   b. $0, 1, 2, \alpha, \alpha+1, \alpha+2, 2\alpha, 2\alpha+1, 2\alpha+2, \alpha^2, \alpha^2+1, \alpha^2+2, 2\alpha^2, 2\alpha^2+1, 2\alpha^2+2, \alpha^2+\alpha, \alpha^2+\alpha+1, \alpha^2+\alpha+2, 2\alpha^2+\alpha, 2\alpha^2+\alpha+1, 2\alpha^2+\alpha+2, \alpha^2+2\alpha, \alpha^2+2\alpha+1, \alpha^2+2\alpha+2, 2\alpha^2+2\alpha, 2\alpha^2+2\alpha+1, 2\alpha^2+2\alpha+2$

8. k^n 9. $\left(-4 + \sqrt[3]{2} - 3\sqrt[3]{4}\right)/22$

11. a. $3, 4$ b. No zero in $\mathbf{Z}_5$ c. $2, 3$ d. No zero in $\mathbf{Z}_7$ e. $5, 5$ f. $5, 6$

12. a. $c = 3$ b. 1 13. $\alpha, 2\alpha + 1$ 14. α, $2\alpha + 2$

15. α, $2\alpha^2 + 3\alpha$, $3\alpha^2 + \alpha + 4$ 16. α, $\alpha^2 + 4\alpha$, $4\alpha^2 + 3$

Appendix Exercises

1. For $x = 0$, the statement $0^2 > 0$ is false.

2. For $x = \frac{1}{2}$, the statement $x^2 \geq x$ is false.

3. For $a = 0$ and any real number b, the statement $0 \cdot b = 1$ is false.

4. For $x = -1$, the statement $2^x < 3^x$ is false.

5. For $x = -4$, the statement $-(-4) < |-4|$ is false.

6. For the real number $x = 0$, it is true that $0 < 1$, but $0^2 < 0$ is false.

7. For $n = 6$, the statement $6^2 + 2(6) = 48$ is true.

8. For $x = \frac{2}{3}$, the statement $\frac{2}{3} + \frac{1}{\frac{2}{3}} = \frac{13}{6}$ is true.

9. For $n = 5$, the statement $5^2 < 2^5$ is true.

10. For $n = 4$, the statement $1 + 3(4) < 2^4$ is true.

11. For $n = 3$, the integer $3^2 + 3$ is an even integer.

12. For $n = 3$, the integer $3^2 + 2(3)$ is a multiple of 5.

13. There is at least one child who did not receive a Valentine card.

14. There is at least one house that does not have a fireplace.

15. There is at least one senior who either did not graduate or did not receive a job offer.

16. There is at least one cheerleader who either is not tall or is not athletic.

17. All of the apples in the basket are not rotten.

18. All the snakes are poisonous.

19. All of the politicians are dishonest or untrustworthy.

20. All the cold medications are unsafe or ineffective.

21. There is at least one $x \in A$ such that $x \notin B$.

22. There exists a real number r such that the square of r is negative.

23. There exists a right triangle with sides a and b and hypotenuse c such that $c^2 \neq a^2 + b^2$.

24. There exist two rational numbers r and s such that there is no irrational number j between them.

25. Some complex number does not have a multiplicative inverse.

26. There exist two 2×2 matrices A and B over the real numbers such that $AB \neq BA$.

27. There are sets A and B such that the Cartesian products $A \times B$ and $B \times A$ are not equal.

28. There exists a real number c such that $x < y \nRightarrow cx < cy$.

29. For every complex number $x, x^2 + 1 \neq 0$.

30. For every 2×2 matrix A over the real numbers, $A^2 \neq I$.

31. For all sets A and B, the set A is not a subset of $A \cap B$.

32. For every complex number z, $\overline{z} \neq z$.

33. For any triangle with angles α, β, and γ, the inequality $\alpha + \beta + \gamma \leq 180°$ holds.

34. For every angle $\theta, \sin\theta \neq 2.1$.

35. For every real number $x, 2^x > 0$.

36. For every even integer x, x^2 is even.

37. TRUTH TABLE for $p \Leftrightarrow \sim (\sim p)$

p	$\sim p$	$\sim (\sim p)$
T	F	T
F	T	F

We examine the two columns headed by p and $\sim (\sim p)$ and note that they are identical.

38. TRUTH TABLE for $p \vee (\sim p)$

p	$\sim p$	$p \vee (\sim p)$
T	F	T
F	T	T

39. TRUTH TABLE for $\sim (p \wedge (\sim p))$

p	$\sim p$	$p \wedge (\sim p)$	$\sim (p \wedge (\sim p))$
T	F	F	T
F	T	F	T

40. TRUTH TABLE for $p \Rightarrow (p \vee q)$

p	q	$p \vee q$	$p \Rightarrow (p \vee q)$
T	T	T	T
T	F	T	T
F	T	T	T
F	F	F	T

41. TRUTH TABLE for $(p \wedge q) \Rightarrow p$

p	q	$p \wedge q$	$(p \wedge q) \Rightarrow p$
T	T	T	T
T	F	F	T
F	T	F	T
F	F	F	T

42. TRUTH TABLE for $\sim(p \vee q) \Leftrightarrow (\sim p) \wedge (\sim q)$

p	q	$p \vee q$	$\sim(p \vee q)$	$\sim p$	$\sim q$	$(\sim p) \wedge (\sim q)$
T	T	T	F	F	F	F
T	F	T	F	F	T	F
F	T	T	F	T	F	F
F	F	F	T	T	T	T

We examine the two columns headed by $\sim(p \vee q)$ and $(\sim p) \wedge (\sim q)$ and note that they are identical.

43. TRUTH TABLE for $(p \wedge (p \Rightarrow q)) \Rightarrow q$

p	q	$p \Rightarrow q$	$p \wedge (p \Rightarrow q)$	$(p \wedge (p \Rightarrow q)) \Rightarrow q$
T	T	T	T	T
T	F	F	F	T
F	T	T	F	T
F	F	T	F	T

44. TRUTH TABLE for $(p \Rightarrow q) \Leftrightarrow \sim(p \wedge \sim q)$

p	q	$p \Rightarrow q$	$\sim q$	$p \wedge \sim q$	$\sim(p \wedge \sim q)$
T	T	T	F	F	T
T	F	F	T	T	F
F	T	T	F	F	T
F	F	T	T	F	T

We examine the two columns headed by $p \Rightarrow q$ and $\sim(p \wedge \sim q)$ and note that they are identical.

45. TRUTH TABLE for $(p \Rightarrow q) \Leftrightarrow ((\sim p) \vee q)$

p	q	$p \Rightarrow q$	$\sim p$	$(\sim p) \vee q$
T	T	T	F	T
T	F	F	F	F
F	T	T	T	T
F	F	T	T	T

We examine the two columns headed by $p \Rightarrow q$ and $(\sim p) \vee q$ and note that they are identical.

46. TRUTH TABLE for $(\sim (p \Rightarrow q)) \Leftrightarrow (p \wedge (\sim q))$

p	q	$p \Rightarrow q$	$\sim (p \Rightarrow q)$	$\sim q$	$p \wedge (\sim q)$
T	T	T	F	F	F
T	F	F	T	T	T
F	T	T	F	F	F
F	F	T	F	T	F

We examine the two columns headed by $\sim (p \Rightarrow q)$ and $p \wedge (\sim q)$ and note that they are identical.

47. TRUTH TABLE for $(p \Rightarrow q) \Leftrightarrow ((p \wedge (\sim q)) \Rightarrow (\sim p))$

p	q	$p \Rightarrow q$	$\sim q$	$p \wedge (\sim q)$	$\sim p$	$(p \wedge (\sim q)) \Rightarrow (\sim p)$
T	T	T	F	F	F	T
T	F	F	T	T	F	F
F	T	T	F	F	T	T
F	F	T	T	F	T	T

We examine the two columns headed by $p \Rightarrow q$ and $(p \wedge (\sim q)) \Rightarrow (\sim p)$ and note that they are identical.

48. TRUTH TABLE for $r \vee (p \wedge q) \Leftrightarrow (r \vee p) \wedge (r \vee q)$

p	q	r	$p \wedge q$	$r \vee (p \wedge q)$	$r \vee p$	$r \vee q$	$(r \vee p) \wedge (r \vee q)$
T	T	T	T	T	T	T	T
T	T	F	T	T	T	T	T
T	F	T	F	T	T	T	T
T	F	F	F	F	T	F	F
F	T	T	F	T	T	T	T
F	T	F	F	F	F	T	F
F	F	T	F	T	T	T	T
F	F	F	F	F	F	F	F

We examine the two columns headed by $r \vee (p \wedge q)$ and $(r \vee p) \wedge (r \vee q)$ and note that they are identical.

49. TRUTH TABLE for $(p \wedge q \wedge r) \Rightarrow ((p \vee q) \wedge r)$

p	q	r	$p \wedge q \wedge r$	$p \vee q$	$(p \vee q) \wedge r$	$(p \wedge q \wedge r) \Rightarrow ((p \vee q) \wedge r)$
T	T	T	T	T	T	T
T	T	F	F	T	F	T
T	F	T	F	T	T	T
T	F	F	F	T	F	T
F	T	T	F	T	T	T
F	T	F	F	T	F	T
F	F	T	F	F	F	T
F	F	F	F	F	F	T

50. TRUTH TABLE for $((p \Rightarrow q) \wedge (q \Rightarrow r)) \Rightarrow (p \Rightarrow r)$

p	q	r	$p \Rightarrow q$	$q \Rightarrow r$	$(p \Rightarrow q) \wedge (q \Rightarrow r)$	$p \Rightarrow r$	$((p \Rightarrow q) \wedge (q \Rightarrow r)) \Rightarrow (p \Rightarrow r)$
T	T	T	T	T	T	T	T
T	T	F	T	F	F	F	T
T	F	T	F	T	F	T	T
T	F	F	F	T	F	F	T
F	T	T	T	T	T	T	T
F	T	F	T	F	F	T	T
F	F	T	T	T	T	T	T
F	F	F	T	T	T	T	T

51. TRUTH TABLE for $(p \Rightarrow (q \wedge r)) \Leftrightarrow ((p \Rightarrow q) \wedge (p \Rightarrow r))$

p	q	r	$q \wedge r$	$p \Rightarrow (q \wedge r)$	$p \Rightarrow q$	$p \Rightarrow r$	$(p \Rightarrow q) \wedge (p \Rightarrow r)$
T	T	T	T	T	T	T	T
T	T	F	F	F	T	F	F
T	F	T	F	F	F	T	F
T	F	F	F	F	F	F	F
F	T	T	T	T	T	T	T
F	T	F	F	T	T	T	T
F	F	T	F	T	T	T	T
F	F	F	F	T	T	T	T

We examine the two columns headed by $p \Rightarrow (q \wedge r)$ and $(p \Rightarrow q) \wedge (p \Rightarrow r)$ and note that they are identical.

52. TRUTH TABLE for $((p \wedge q) \Rightarrow r) \Leftrightarrow (p \Rightarrow (q \Rightarrow r))$

p	q	r	$p \wedge q$	$(p \wedge q) \Rightarrow r$	$q \Rightarrow r$	$p \Rightarrow (q \Rightarrow r)$
T	T	T	T	T	T	T
T	T	F	T	F	F	F
T	F	T	F	T	T	T
T	F	F	F	T	T	T
F	T	T	F	T	T	T
F	T	F	F	T	F	T
F	F	T	F	T	T	T
F	F	F	F	T	T	T

We examine the two columns headed by $(p \wedge q) \Rightarrow r$ and $p \Rightarrow (q \Rightarrow r)$ and note that they are identical.

53. The *implication* $(p \Rightarrow q)$ is true: My grade for this course is A implies that I can enroll in the next course.

The *contrapositive* $(\sim q \Rightarrow \sim p)$ is true: I cannot enroll in the next course implies that my grade for this course is not A.

The *inverse* $(\sim p \Rightarrow \sim q)$ is false: My grade for this course in not A implies that I cannot enroll in the next course.

The *converse* $(q \Rightarrow p)$ is false: I can enroll in the next course implies that my grade for this course is A.

54. The *implication* $(p \Rightarrow q)$ is true: My car ran out of gas implies my car won't start.

The *contrapositive* $(\sim q \Rightarrow \sim p)$ is true: My car will start implies my car did not run out of gas.

The *inverse* $(\sim p \Rightarrow \sim q)$ is false: My car did not run out of gas implies that my car will start.

The *converse* $(q \Rightarrow p)$ is false: My car won't start implies my car ran out of gas.

55. The *implication* $(p \Rightarrow q)$ is true: The Saints win the Super Bowl implies that the Saints are the champion football team.

The *contrapositive* $(\sim q \Rightarrow \sim p)$ is true: The Saints are not the champion football team implies that the Saints did not win the Super Bowl.

The *inverse* $(\sim p \Rightarrow \sim q)$ is true: The Saints did not win the Super Bowl implies that the Saints are not the champion football team.

The *converse* $(q \Rightarrow p)$ is true: The Saints are the champion football team implies that the Saints did win the Super Bowl.

56. The *implication* $(p \Rightarrow q)$ is true: I have completed all the requirements for a bachelor's degree implies I can graduate with a bachelor's degree.

The *contrapositive* $(\sim q \Rightarrow \sim p)$ is true: I cannot graduate with a bachelor's degree implies that I have not completed all the requirements for a bachelor's degree.

The *inverse* $(\sim p \Rightarrow \sim q)$ is true: I have not completed all the requirements for a bachelor's degree implies I cannot graduate with a bachelor's degree.

The *converse* $(q \Rightarrow p)$ is true: I can graduate with a bachelor's degree implies I have completed all the requirements for a bachelor's degree.

57. The *implication* $(p \Rightarrow q)$ is false: My pet has four legs implies that my pet is a dog.

The *contrapositive* $(\sim q \Rightarrow \sim p)$ is false: My pet is not a dog implies that my pet does not have four legs.

The *inverse* $(\sim p \Rightarrow \sim q)$ is true: My pet does not have four legs implies that my pet is not a dog.

The *converse* $(q \Rightarrow p)$ is true: My pet is a dog implies that my pet has four legs.

58. The *implication* $(p \Rightarrow q)$ is false: I am within 30 miles of home implies I am within 20 miles of home.

The *contrapositive* $(\sim q \Rightarrow \sim p)$ is false: I am more than 20 miles from home implies I am more than 30 miles from home.

The *inverse* $(\sim p \Rightarrow \sim q)$ is true: I am more than 30 miles from home implies I am more than 20 miles from home.

The *converse* ($q \Rightarrow p$) is true: I am within 20 miles of home implies I am within 30 miles of home.

59. The *implication* ($p \Rightarrow q$) is true: Quadrilateral $ABCD$ is a square implies that quadrilateral $ABCD$ is a rectangle.

The *contrapositive* ($\sim q \Rightarrow \sim p$) is true: Quadrilateral $ABCD$ is not a rectangle implies that quadrilateral $ABCD$ is not a square.

The *inverse* ($\sim p \Rightarrow \sim q$) is false: Quadrilateral $ABCD$ is not a square implies that quadrilateral $ABCD$ is not a rectangle.

The *converse* ($q \Rightarrow p$) is false: Quadrilateral $ABCD$ is a rectangle implies that quadrilateral $ABCD$ is a square.

60. The *implication* ($p \Rightarrow q$) is false: Triangle ABC is isosceles implies triangle ABC is equilateral.

The *contrapositive* ($\sim q \Rightarrow \sim p$) is false: Triangle ABC is not equilateral implies triangle ABC is not isosceles.

The *inverse* ($\sim p \Rightarrow \sim q$) is true: Triangle ABC is not isosceles implies triangle ABC is not equilateral.

The *converse* ($q \Rightarrow p$) is true: Triangle ABC is equilateral implies triangle ABC is isosceles.

61. The *implication* ($p \Rightarrow q$) is true: x is a positive real number implies that x is a nonnegative real number.

The *contrapositive* ($\sim q \Rightarrow \sim p$) is true: x is a negative real number implies that x is a nonpositive real number.

The *inverse* ($\sim p \Rightarrow \sim q$) is false: x is a nonpositive real number implies that x is a negative real number.

The *converse* ($q \Rightarrow p$) is false: x is a nonnegative real number implies that x is a positive real number.

62. The *implication* ($p \Rightarrow q$) is true: x is a positive real number implies x^2 is a positive real number.

The *contrapositive* ($\sim q \Rightarrow \sim p$) is true: x^2 is not a positive real number implies that x is not a positive real number.

The *inverse* ($\sim p \Rightarrow \sim q$) is false: x is not a positive real number implies x^2 is not a positive real number.

The *converse* ($q \Rightarrow p$) is false: x^2 is a positive real number implies x is a positive real number.

63. The *implication* ($p \Rightarrow q$) is true: $5x$ is odd implies that x is odd.

The *contrapositive* ($\sim q \Rightarrow \sim p$) is true: x is not odd implies that $5x$ is not odd.

The *inverse* ($\sim p \Rightarrow \sim q$) is true: $5x$ is not odd implies that x is not odd.

The *converse* ($q \Rightarrow p$) is true: x is odd implies that $5x$ is odd.

64. The *implication* $(p \Rightarrow q)$ is true: $5+x$ is odd implies x is even.

The *contrapositive* $(\sim q \Rightarrow \sim p)$ is true: x is odd implies $5+x$ is even.

The *inverse* $(\sim p \Rightarrow \sim q)$ is true: $5+x$ is even implies x is odd.

The *converse* $(q \Rightarrow p)$ is true: x is even implies $5+x$ is odd.

65. The *implication* $(p \Rightarrow q)$ is true: xy is even implies that x is even or y is even.

The *contrapositive* $(\sim q \Rightarrow \sim p)$ is true: x is odd and y is odd implies that xy is odd.

The *inverse* $(\sim p \Rightarrow \sim q)$ is true: xy is odd implies that x is odd and y is odd.

The *converse* $(q \Rightarrow p)$ is true: x is even or y is even implies that xy is even.

66. The *implication* $(p \Rightarrow q)$ is true: x is even and y is even implies $x+y$ is even.

The *contrapositive* $(\sim q \Rightarrow \sim p)$ is true: $x+y$ is odd implies at least one of x or y is odd.

The *inverse* $(\sim p \Rightarrow \sim q)$ is false: At least one of x or y is odd implies $x+y$ is odd.

The *converse* $(q \Rightarrow p)$ is false: $x+y$ is even implies x is even and y is even.

67. The *implication* $(p \Rightarrow q)$ is false: $x^2 > y^2$ implies that $x > y$.

The *contrapositive* $(\sim q \Rightarrow \sim p)$ is false: $x \leq y$ implies that $x^2 \leq y^2$.

The *inverse* $(\sim p \Rightarrow \sim q)$ is false: $x^2 \leq y^2$ implies that $x \leq y$.

The *converse* $(q \Rightarrow p)$ is false: $x > y$ implies that $x^2 > y^2$.

68. The *implication* $(p \Rightarrow q)$ is true: $\frac{x}{y} > 0$ implies $xy > 0$.

The *contrapositive* $(\sim q \Rightarrow \sim p)$ is true: $xy \leq 0$ implies $\frac{x}{y} \leq 0$ or $\frac{x}{y}$ is not defined.

The *inverse* $(\sim p \Rightarrow \sim q)$ is true: $\frac{x}{y}$ is undefined or $\frac{x}{y} \leq 0$ implies that $xy \leq 0$.

The *converse* $(q \Rightarrow p)$ is true: $xy > 0$ implies $\frac{x}{y} > 0$.

69. Contrapositive: $\sim (q \vee r) \Rightarrow \sim p$, or $((\sim q) \wedge (\sim r)) \Rightarrow \sim p$

Converse: $(q \vee r) \Rightarrow p$

Inverse: $\sim p \Rightarrow \sim (q \vee r)$, or $\sim p \Rightarrow ((\sim q) \wedge (\sim r))$

70. Contrapositive: $\sim (q \wedge r) \Rightarrow \sim p$, or $((\sim q) \vee (\sim r)) \Rightarrow \sim p$

Converse: $(q \wedge r) \Rightarrow p$

Inverse: $\sim p \Rightarrow \sim (q \wedge r)$, or $\sim p \Rightarrow ((\sim q) \vee (\sim r))$

71. Contrapositive: $q \Rightarrow \sim p$

Converse: $\sim q \Rightarrow p$

Inverse: $\sim p \Rightarrow q$

72. Contrapositive: $\sim(\sim p) \Rightarrow \sim(p \wedge \sim q)$, or $p \Rightarrow ((\sim p) \vee q)$

Converse: $\sim p \Rightarrow (p \wedge \sim q)$

Inverse: $\sim(p \wedge \sim q) \Rightarrow \sim(\sim p)$, or $((\sim p) \vee q) \Rightarrow p$

73. Contrapositive: $\sim(r \wedge s) \Rightarrow \sim(p \vee q)$, or $((\sim r) \vee (\sim s)) \Rightarrow ((\sim p) \wedge (\sim q))$

Converse: $(r \wedge s) \Rightarrow (p \vee q)$

Inverse: $\sim(p \vee q) \Rightarrow \sim(r \wedge s)$, or $((\sim p) \wedge (\sim q)) \Rightarrow ((\sim r) \vee (\sim s))$

74. Contrapositive: $\sim(r \wedge s) \Rightarrow \sim(p \wedge q)$, or $((\sim r) \vee (\sim s)) \Rightarrow ((\sim p) \vee (\sim q))$

Converse: $(r \wedge s) \Rightarrow (p \wedge q)$

Inverse: $\sim(p \wedge q) \Rightarrow \sim(r \wedge s)$, or $((\sim p) \vee (\sim q)) \Rightarrow ((\sim r) \vee (\sim s))$